W9-AAU-463

THE CALL OF ANTARCTICA

EXPLORING AND PROTECTING EARTH'S COLDEST CONTINENT

LEILANI RAASHIDA HENRY

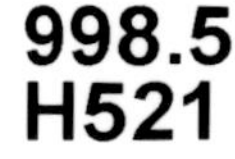

TWENTY-FIRST CENTURY BOOKS / MINNEAPOLIS

THIS BOOK IS DEDICATED TO A HEALTHY PLANET, HEALTHY PEOPLE, AND A HEALTHY CONNECTION BETWEEN EARTH AND ALL ITS BEINGS.

THANK YOU TO MY FRIENDS, FAMILY, AND COLLEAGUES WHO SUPPORTED AND ENDURED MY YEARS OF FOCUS ON THIS PROJECT. A SPECIAL THANKS TO DR. TED SCAMBOS, ANTARCTIC SCIENTIST, WHO HELPED MAKE THIS BOOK POSSIBLE.

Twenty-First Century Books™
An imprint of Lerner Publishing Group, Inc.
241 First Avenue North
Minneapolis, MN 55401 USA

For reading levels and more information, look up this title at www.lernerbooks.com.

Main body text set in Adobe Garamond Pro Regular.
Typeface provided by Adobe Systems.

Library of Congress Cataloging-in-Publication Data

Names: Henry, Leilani Raashida, author.
Title: The call of Antarctica : exploring and protecting earth's coldest continent / Leilani Raashida Henry.
Description: Minneapolis, MN : Twenty-First Century Books, [2022] | Includes bibliographical references and index. | Audience: Ages 13–18 | Audience: Grades 10–12 | Summary: "Author Leilani Raashida Henry, daughter of the first person of African descent to go to Antarctica, recounts her father's expedition while educating on the incredible geography, biodiversity, and history of the continent"— Provided by publisher.
Identifiers: LCCN 2020044845 (print) | LCCN 2020044846 (ebook) | ISBN 9781541560956 (library binding) | ISBN 9781728401607 (ebook)
Subjects: LCSH: Antarctica—Juvenile literature. | Antarctica—Discovery and exploration—Juvenile literature. | Biodiversity—Antarctica—Juvenile literature.
Classification: LCC G863 .H46 2022 (print) | LCC G863 (ebook) | DDC 919.89—dc23

LC record available at https://lccn.loc.gov/2020044845
LC ebook record available at https://lccn.loc.gov/2020044846

Manufactured in the United States of America
1-46395-47485-6/10/2021

CONTENTS

FOREWORD BY DR. TED SCAMBOS

The call of Antarctica is a call to discovery. When you begin to learn about the continent, you sense that it is very different from any place you've been before. Antarctica is a part of Earth, and yet it's unearthly—a vast, ice-covered land where all the rules are different. For many, the strangeness of it all inspires them to go there to see it firsthand, and often one visit is not enough. The adventure of traveling there, the landscape, and fantastic ice structures create excitement, a sense of adventure, and a deep appreciation of the scale, the silence, and the grandeur of the southernmost land. No matter how you visit the ice sheet—sailing along its coast, on a research expedition, striving to reach the continent's farthest points, or simply working at a research station's galley—the feeling of wanting to see it again is the same. Even when explorers' lives were threatened by the extreme conditions and hardships, they wanted to return.

Antarctica also inspires a more personal discovery, one that you don't anticipate. Whether you go as part of a team of researchers or with a group of excited tourists, you will find not only a unique, beautiful place but also something new about yourself. The scale of Antarctica—its silence, the beauty of its natural landscape—lead you inward as well as onward. On a scientific expedition, the relentless work and the close social interaction of the field team test you. Isolation and the many other challenges that accompany exploration are all part of an internal survey as well. You discover new things about who you are, how far you can reach, and how you fit in. Even if you are a solo adventurer, you encounter the same questions. You ask yourself, "How can *I* be a better part of *we*?"

This book is a brief introduction to the continent of Antarctica and its surrounding ocean, an overview of its history, and a sample of the science behind how the continent works. It touches upon Antarctica's geology, biology, and environment. As a way of discovering more about the continent, we follow the story of a member of a past Antarctic expedition—the American George W. Gibbs Jr. In 1940, at twenty-three years of age, Gibbs was the first Black man to set foot on the continent.

Gibbs experienced both the inner and outer kinds of discovery on his journey. He had the adventure of his lifetime, and he embraced his role as a part of a diverse team. He saw landscapes and wildlife that awed him. He felt the burn of racial discrimination, but he also knew interracial teamwork and harmony. He saw how other nations put less importance on skin color and more importance on fellowship. These experiences shaped him and set him on a course to become a civic leader and civil rights advocate in his state and nation. Above all, his example shows the power of outlook, attitude, and optimism. No matter the situation, Gibbs steered clear of resentment and focused on experiences and on being present. He appreciated every aspect of his journey, inward and outward, and spoke about the experience for the rest of his life.

He felt the call of Antarctica.

Ted Scambos is a senior research scientist at the University of Colorado in Boulder. He has been on twenty expeditions to Antarctica and the Southern Ocean to study the ice and the impacts of climate change.

INTRODUCTION
INTO THE ICE

The ship bucks like a wild horse. To the crew on board, it seems the bucking will never end. It's impossible to walk on deck without holding on or falling. The fog is so thick that the men can't see their hands in front of their faces. When the grinding of the ship's engine stops, the silence is startling.

Forty US Navy crew and twenty-two civilians are aboard the USS *Bear*, a three-mast, sixty-six-year-old wooden sailing ship, modified with a new diesel engine and steel sheath to protect the hull from the ice. The fog is so thick that the men can hardly see their hands in front of their faces. The water is about 31°F (–0.5°C)—just above the freezing point for salt water. The freezing spray covers everything and makes the deck slippery underfoot. They haven't had a warm, cooked meal in days, only sandwiches. The seas have been too rough for the galley. Gray, white, and blue are the colors of the landscape—reflected in the sky, the clouds, the salt water, and the fog.

Days grow longer and longer as the nights shrink to nothing—but this doesn't trouble the crew. Everyone aboard the US Navy ship is too excited to sleep anyway. The purple, pink, and tangerine sunrises—actually sunsets that never quite finish—periodically fill the sky above an endless indigo-blue sea. The sun's halo is a pastel rainbow. But now, moving slowly and carefully into the ice pack, an eerie calmness takes over.

The grueling work of bashing through the ice is not over, but the much-anticipated glimpse of the first iceberg is here. The huge

Voyagers to Antarctica are treated to a landscape surprisingly full of colors.

ice mountain has sharp, otherworldly facets and edges that reflect the sunlight at different angles as the ship glides by. Members of the crew see familiar objects in the iceberg's shape: a huge yacht, a castle, a cityscape, an island with a lighthouse. For miles, the ship passes towering hunks of blue, green, and pink ice. The ice cliffs soar as high as ten stories.

It is 1939. The *Bear* has spent almost thirty-nine straight days at sea, far from any land, and has begun plowing through a flat

Explorers arriving in Antarctica for the first time are often awed by huge icebergs that range in size, shape, and color. The largest can span hundreds of square miles, larger than many cities.

expanse of ice on its way to the southernmost part of the world. The *Bear*'s newly built sister ship, the *North Star*, left one week earlier. It's waiting for the *Bear* at the Bay of Whales, the southernmost point a ship can reach. The *Bear* fights its way in a zigzag pattern through the ice pack. When the ship can't move on, it creates holes in the ice or finds openings in the pack—sometimes banging into the ice head-on. Bone-chilling winds blow over the icy water. The deep green sea churns up white foam among the bright blue ice. The icebergs, which earlier in the voyage towered above the water like jagged rocks, are now mostly flat.

The USS *Bear* was built in 1874 to be a whaling and sealing ship (a ship used to hunt those animals). After the US Navy purchased it in 1884, the ship served on cold-weather rescues and expeditions for decades, including Admiral Richard E. Byrd's in 1939.

Killer whales have been following the ship for a few days. Now the men see Adélie penguins diving in and out of the water on nearby ice floes. A variety of seabirds—petrels, Cape pigeons, skuas, and albatross—let the expedition members know that they are almost to their destination: Antarctica.

CHAPTER 1
UNKNOWN SOUTHERN LAND

TODAY AT NOON, EN ROUTE TO LITTLE AMERICA, WE ATE OUR FIRST MEAL SOUTH OF CAPE HORN. IF CAPE HORN IS AS ROUGH AS THIS SEA IS TONIGHT, I AM NOT AS ANXIOUS TO GO AROUND IT AS I WAS. I CAN HARDLY WRITE, AS WE ARE IN A SEA, THAT IS A SEA. IF YOU DON'T KNOW WHAT I MEAN, TRY BREAKING A WILD HORSE IN. OF COURSE, THE HORSE WOULD GET TIRED AND GIVE IN, BUT THIS IS GETTING TO BE A 24-HOUR ROUTINE. WE HAD A HEAVY SLEET THIS MORNING. IT'S VERY COLD AND THE BOYS ARE PUTTING HEAVY UNDERWEAR ON.

—George W. Gibbs Jr. aboard the USS *Bear*, December 27, 1939

Antarctica, the fifth-largest continent on Earth, is the coldest, windiest, highest, driest, and most remote part of the world. It's the southernmost land on Earth, with the South Pole at its center. Temperatures there are bitterly cold, with readings plunging as low as

–140°F (–95.5°C) in winter. Even in summer, temperatures only rise above 32°F (0°C) in the warmest coastal areas. The winds blowing across Antarctica can be ferocious, clocking in at nearly 200 miles (322 km) per hour near the edge of the ice sheet.

The South Pole tilts away from the sun in winter (June to September in the southern half of the globe). Because of this tilt, almost all of Antarctica is covered in darkness—with no daylight—in wintertime. In summer (December through March), Antarctica tilts toward the sun, meaning that the sun doesn't set for months at a time. Sometimes Antarctica is said to have only two seasons, summer and winter, because it has about six months of darkness and then six months of daylight.

The continent covers 5.5 million square miles (14 million sq. km). That's larger than the United States and Mexico combined. Thousands of rocky islands line its coasts. The continent is surrounded by the Southern Ocean, a vast expanse of water comprised of the southernmost reaches of the Pacific, Atlantic, and Indian Oceans. Also known as the Antarctic Ocean, the Southern Ocean is full of icebergs. It's also the roughest, harshest sea on the planet. Powerful winds from the west drive towering waves, sometimes more than 30 feet (9 m) high. These same winds create currents that propel huge volumes of water, the largest in the world.

Most of Antarctica is covered by a massive ice sheet, an enormous layer of ice that flows slowly across the land. Antarctica's ice sheet is divided into two sections, one in the Eastern Hemisphere of the continent (the side near Africa, India, and Australia) and one in the Western Hemisphere (south of the Pacific Ocean and South America). The Antarctic Peninsula, south of Cape Horn in South America, is home to a coast of fjords, rocky islands, and hundreds of glaciers. The miles-thick ice sheet and myriad Antarctic glaciers hold about 90 percent of the world's ice and 70 percent of its fresh water.

The word Antarctica comes from the Greek word *antarktike*, meaning "opposite of the Arctic." The Arctic is the region surrounding

The Ross Ice Shelf was one of the first parts of Antarctica encountered by human beings. It covers an area of the ocean nearly the same size as France.

Earth's North Pole—the very top of the world. Antarctica is its opposite, surrounding the pole at the very bottom. An ancient Greek geographer Marinus of Tyre gave Antarctica its name.

Ancient peoples never visited Antarctica. But ancient geographers suspected it existed because they believed that the way Earth achieved balance was for each landmass to have a complementary landmass on the opposite side of the globe. The Arctic had a landmass surrounded by an ocean in the north, and so from a very early time, cartographers depicted a large landmass surrounded by an ocean in the south. In 150 CE, the Greek geographer Ptolemy called this region Terra Australis Incognita, meaning "unknown southern land."

Over the following centuries, European explorers learned more about Earth's geography. From Europe, ships sailed south, around the southern tip of Africa and then east to Asia. They traveled west to the Americas. The explorers mapped the lands and bodies of water they encountered. Cartographers created maps of the world, such as the Piri Reis map of 1513, the Orontius Finaeus map of 1532, and the Mercator map of 1569. All these maps show a continent at the bottom of the world, even though no one knew for sure that it existed.

HEADING SOUTH

A legend says that a chief of the Māori people, the Indigenous inhabitants of New Zealand, first saw Antarctica on a sea voyage in the seventh century CE. In 1772 the British government sent naval captain James Cook to search for the land that was shown on so many old maps: Terra Australis Incognita. With his ships *Resolution* and *Adventure*, Cook was the first to cross the Antarctic Circle, an imaginary line around Earth that encloses most of Antarctica. In 1773 he came within 75 miles (120 km) of the continent, but ice and bitter cold forced him to turn around before seeing it. Even so, Cook concluded that the southern continent, if it existed, would likely be covered in ice. He said it would be far too cold for farming and human settlement.

In the early 1800s, more Europeans sailed through the Southern Ocean. Rather than explorers, they were whale and seal hunters. They found the animals they hunted living around the islands of the Antarctic Peninsula. They hunted whales for their valuable blubber and baleen, a hornlike substance found in whale jaws, and seals for their valuable skins.

In 1819 the Russian government sent Captain Fabian Gottlieb von Bellingshausen on a voyage around the world. With his ships the *Vostok* and *Mirny*, he reported a location that was within 20 miles (32 km) of the Antarctic Peninsula on January 28, 1820. Only a few days later, British sea captain Edward Bransfield also visited the Antarctic Peninsula. Later that year, on November 18, twenty-one-year-old Nathaniel Brown Palmer, captain of the American sealing sloop *Hero*, recorded sighting the peninsula.

Expeditions kept coming. Scotsman James Weddell was a commercial sealer, explorer, and captain in the British Royal Navy. In 1823 he discovered the Weddell Sea, which borders the Antarctic Peninsula. The sea and Weddell seals are named after him.

In 1838, wanting to keep up with its European rivals, the United

States sent a military expedition to Antarctica. Naval officer Charles Wilkes sailed with a fleet of six ships on a voyage that lasted two years. His sighting of the extensive coast far to the south of Australia proved that Antarctica was a vast continent.

Jules Sébastien César Dumont d'Urville's expedition in 1840 was sponsored by the French government. D'Urville's ships were the *Astrolabe* and the smaller *Zélée*. He spotted the East Antarctic coastline, but pack ice blocked his way to the land. His voyage was further hampered when much of his crew got sick with scurvy, a disease caused by a lack of vitamin C in the diet. D'Urville left his mark on Antarctica by naming Adélie penguins and a region called Adélie Land for his wife.

James Clark Ross led a British expedition with two warships in 1839. He discovered a large gulf he named the Ross Sea and was the first to see the ice barrier, a great wall of ice that blocked further progress to the southern part of the sea. It was later named the Ross Ice Shelf in his honor. His expedition was also the first to see a region called Victoria Land, named for Britain's queen at the time. While on the newly discovered Ross Island, a blacksmith aboard the *Erebus*, one of Ross's ships, spotted a volcano. He called it a "splendid Burning Mountain." Ross later named the volcano Mount Erebus after the ship. In addition to these geographic discoveries, Ross observed the local wildlife. He saw how penguins use pebbles to build their nests and watched them regurgitate food into their chicks' mouths to feed them.

In these days before photography, most sea expeditions carried an artist to document landscapes, life on the ships, and the plants and animals discovered during expeditions. John Davis, an officer on Ross's other ship *Terror*, was the artist of the expedition. He made paintings of Mount Erebus, the Ross Ice Shelf, and other scenes.

Carl Anton Larsen led the first Norwegian expedition to Antarctica from 1892 to 1894. This expedition brought back the first fossils from

A gentoo penguin feeds its chick. All species of penguins feed their young using some form of regurgitation.

the continent. These were ancient bones that had turned to stone or rock layers that showed impressions of ancient plant and animal life. Another Norwegian ship, *Antarctic*, reached the opposite side of the continent on January 24, 1895. Some of the crew went ashore to collect rock samples on Cape Adare.

Ship after ship continued to arrive. These explorations helped to fill in the map of the Antarctic coastline and added to scientific knowledge. Expedition artists continued to capture the splendor of the southern continent. In January 1902, artist Frank Wilbert Stokes visited the Antarctic Peninsula with a Swedish expedition. He is credited with the first painting of an iceberg. In addition to his visual art, he described what he saw. He wrote of "bergs . . . glistening in

While there were many other expeditions that happened on the continent, this map shows the paths of some of the most historic.

marvelous pink purity under the sun's rays, with rich, deep shadows of turquoise-cobalt blue," "whirling cloud-masses of dark smoky blue," and "the glaciers' brilliance of pale purple and gold."

BOOTS ON THE GROUND

Though a few explorers had gone ashore on the Antarctic mainland, no one had yet made a journey deep inland. That changed in November 1902 with an expedition led by British naval captain Robert Falcon Scott. With a crew of explorers and scientists, he sailed in a ship called *Discovery* to the Ross Ice Shelf. Leaving most of his men at the ship, Scott, accompanied by crew members Ernest Shackleton and Edward Wilson, disembarked and headed south across the ice shelf. They loaded gear onto sledges, or sleds, which were pulled by huskies.

At camp, Ernest Shackleton (*right*) rests while fellow expedition member Frank Wild skins a fish.

These fast and powerful dogs have coats of thick, warm fur. They provided the pulling power and companionship on this and many later Antarctic expeditions. Arctic Indigenous tribes such as the Inuits had taught white explorers how to travel and survive the cold with huskies.

The three-month-long round trip was treacherous. The huskies grew weak from bad food, and the men shot some of them as food for the other dogs. Without enough dog power, the men had to pull the sledges themselves. On the return leg, Shackleton grew so sick from scurvy that he had to be pulled on the sledge with the gear. The team barely made it back to the ship alive.

Before, Antarctic explorers had one goal in mind: discovering the fabled southern continent and mapping its coast. After Scott's first expedition, the quest was to reach the farthest points on Earth. The geographic South Pole is a fixed point, the meeting place of all the planet's longitudinal lines (imaginary lines that run north-south along

Huskies were frequently used in Antarctic exploration until 1994 when the Antarctic Treaty required them to be removed from the continent to protect native species.

Earth's surface), and the place where Earth's axis of rotation intersects its surface.

After Scott's *Discovery* expedition, Ernest Shackleton was determined to return to the continent to reach the South Pole and the magnetic south pole. The magnetic south pole is determined by Earth's magnetic field, an energy field that makes Earth act like a giant magnet. This pole is constantly moving, influenced by electric currents in Earth's interior.

Sailing aboard the ship *Nimrod*, Shackleton led a sea journey back to Antarctica. Shackleton and three other men headed for the geographic South Pole but ran out of food along the way. They turned back on January 9, 1909, 98 nautical miles (113 miles, or 181 km) short of their goal. Another team of three of his crew, pulling their supplies on sledges, set out across the ice and reached what they thought was the magnetic pole on January 16, 1909. Later explorers determined that this team had miscalculated the pole's location and hadn't actually reached it. Still, the expedition was considered a huge success, mapping large parts of the continent that had never been seen before and briefly setting a record for the highest latitude reached by an expedition to either pole. US Naval officer Robert Peary and explorer Matthew Henson reached the vicinity of the North Pole a few months after Shackleton's farthest journey south.

Robert Scott returned to Antarctica in early 1911 with the ship *Terra Nova*. He too sought the geographic South Pole but also intended to conduct additional scientific and geographic surveys. Before the *Terra Nova* left Australia for New Zealand and then Antarctica, Scott learned that Norwegian explorer Roald Amundsen was also on his way to Antarctica and the pole. The race was on.

Amundsen had an experienced polar team of nineteen men. They had top-notch gear for their attempt to reach the pole: warm sealskin clothing, reindeer fur for sleeping bags, plentiful food, and more than one hundred dogs to pull sledges.

STANDING AT THE NORTH POLE

One of George W. Gibbs's heroes was Matthew Henson (*right*). Born in 1866, Henson belonged to the first generation of free Black people in the United States. He stood at the top of the planet on April 6, 1909. He planted his feet and the American flag at the North Pole as part of Peary's fourth and final attempt at reaching it.

Henson was born to a free Black family in Maryland. He was orphaned at an early age and raised by an abusive stepmother. At the age of twelve, he left home and found his way to the Baltimore shipyard. There he met Captain Childs, an experienced sea captain who hired him as a cabin boy on the *Katie Hines*. Henson spent his teenage years learning a variety of skills on the ship.

By 1887 Henson was an accomplished sailor. He met Robert Peary in Washington, DC, while working as a sales clerk in a furrier shop. When Peary learned of Henson's life and experiences, he proposed a partnership and hired Henson as his personal assistant. Henson's first job with Peary was an expedition to the jungles of Nicaragua to survey a route for a possible canal. Eventually Peary's interest turned to reaching the North Pole. Henson accompanied him on each of his attempts, in 1886, 1898, and 1905. Each time they reached farther north before having to turn back.

Henson came to learn the Inuit language and culture and studied their methods for traveling and survival. He was called Mahri Pahluk, "Matthew the kind one," and became close with an Inuit family. Henson's skills as a sailor, dog handler, sledge driver, craftsperson, fisher, and hunter made him an invaluable partner to Peary.

Henson and Peary made their fourth attempt to reach the North Pole in 1909. They traveled the last leg across the sea ice with four Inuit companions: Ooqueah, Ootah, Egingwah, and Seegloo. On the final day, Henson took the lead, walking ahead of the team. He had estimated the distance left to the pole the night before. After several hours, he stopped and waited for Peary, who was forty-five minutes behind him. He told Peary he thought they had reached the pole or had already passed it. Henson noted in his diary that, after Peary determined their location, it was clear that Henson had passed very near the location of the pole. He told Peary he was first to the pole.

Most now believe that Henson and Peary stood within a few miles of the North Pole on April 6, 1909. But Peary refused to give Henson the honor of "first." He even stopped talking to Henson on the return trip. Upon their return, Peary received all the credit for the expedition, and Henson's key role was underplayed and nearly forgotten.

Henson wrote a book in 1912. Though it was widely read in the African American community, this did not alter the white perception that Peary was the first to reach the North Pole. However, in the foreword to the book, Peary noted that three races stood at the top of the world in harmony.

As the years passed, there was an increased focus on the exact events surrounding Peary and Henson's final expedition. From the beginning, the achievement was challenged by Frederick Cook, who claimed he reached the pole a year earlier. Cook's claim was met with suspicion almost immediately, but it led to a closer inspection of Peary's claim in turn.

Diaries and notes eventually confirmed Peary's and Henson's achievements. In 2000 the National Geographic Society posthumously awarded its highest award, the Hubbard Medal, to Matthew Henson. Peary received the honor first in 1906.

Scott, not expecting a race, had gear more suited to a research expedition. Although he had dogs, he also brought Siberian ponies and British army mules to pull sledges. The ponies, from the cold Russian region of Siberia, could withstand the bitter polar temperatures. But the expedition set out in November, late spring in Antarctica. Summer brought higher temperatures and soft, melting snow. The ponies didn't travel well over this surface and seemed to suffer from the lack of fresh fodder (livestock feed). Scott also brought three motorized sledges, modified from early automobiles. These too bogged down in the soft snow. In the end, Scott relied on what had worked in his first expedition: human strength. Teams of four or five men each dragged sledges loaded with hundreds of pounds of supplies for more than 1,000 miles (1,610 km).

Amundsen's expert team of dog drivers, dog teams, and cross-country skiers easily won the race. They reached the geographic South Pole on December 14, 1911, more than a month ahead of Scott's man-hauling team. Scott had chosen four team members to accompany him on the final leg to the pole. When they arrived on January 17, 1912, they found that Amundsen had planted a Norwegian flag there, as well as a small tent and a note claiming the prize of first to the pole for Norway.

Devastated, the defeated British team set off on their return trip. But they never made it back. Two of the men were injured on the trail and died. Scott and two others continued a few days more but were caught in an unusually long and powerful blizzard. They set up their tent for shelter but were out of food and fuel to melt snow for water. Soon all three died from a combination of cold and starvation, just 11 miles (18 km) from a large depot of food and supplies.

THE SKY'S THE LIMIT

The twentieth century brought a flood of new technology, which explorers quickly employed to help in Antarctic expeditions. In 1911 Douglas Mawson, an Australian geologist, was the first to use

Lincoln Ellsworth (*left*) and Herbert Hollick-Kenyon (*right*) pictured with *Polar Star*. The pair ran out of fuel but couldn't radio for help because their radio was malfunctioning. They were rescued by the British Royal Research Ship *Discovery* in January 1936 at Little America, an exploration base.

radio communications during an expedition. Mawson also took an airplane—another new invention—to Antarctica. He planned to use the plane to survey the territory from above. But winds at Mawson's main camp, near Commonwealth Bay and Cape Denison, toppled the plane and damaged its wings. His men used the plane's stripped-down fuselage, outfitted with skis, to haul equipment instead.

The first explorer to successfully use an airplane on Antarctica was George Hubert Wilkins, an Australian pilot and photographer. With funding from wealthy American publisher William Randolph Hearst, Wilkins made the first successful Antarctic flight on December 20, 1928. He flew along the coast of Graham Land and Palmer Land, on the tip of the Antarctic Peninsula, surveying much yet unknown territory.

US engineer, pilot, and adventurer Lincoln Ellsworth explored both the Arctic and Antarctica. On a 1935 expedition to Antarctica, Ellsworth and copilot Herbert Hollick-Kenyon made the first transantarctic flight in a plane named *Polar Star*. On this trip, they explored the interior of the continent.

SHACKLETON RETURNS

Ernest Shackleton's third and most famous expedition to Antarctica began in 1914. His goal was to trek completely across the continent, a distance of about 1,800 miles (2,897 km). The trip would take his team from the Weddell Sea to the geographic South Pole and then to McMurdo Sound on the other side of Antarctica.

An oft-repeated story is that Shackleton placed an ad in a London newspaper to find his crew: "MEN WANTED for hazardous journey, small wages, bitter cold, long months of complete darkness, constant danger, safe return doubtful, honour and recognition in case of success." Historians have never found the original ad, but the story goes on to say that five thousand people, mostly men, applied for the job. Whatever the truth, Shackleton carefully selected twenty-six men to cross the icy continent from coast to coast accompanied by ninety-six dogs. A support crew was sent out ahead of time to establish caches of food and fuel for them along the route from the pole to McMurdo Sound.

Shackleton's mission never took place. In fact, the main Shackleton party never set foot on mainland Antarctica. Approaching the continent, Shackleton's ship, *Endurance*, fought its way in through 1,000 miles (1,610 km) of pack ice. In January 1915, the ship became stuck in the ice in the southern Weddell Sea. Cut off from all communication with the outside world, the crew stayed on board the ship and waited. As the thick ice continued to grip the ship, darkness settled over Antarctica for the winter.

Shackleton hoped that when summer came, the ice might open up and free the ship to continue its journey. But that didn't happen. The heavy ice crushed the ship, and on November 21, 1915, *Endurance* sank. The crew established a camp, which Shackleton named Patience Camp, on an ice floe. They scavenged supplies, food, and lifeboats from their crushed vessel. They waited again for the ice to thaw. In March 1916, the ice began to break apart, and the crew set off in the lifeboats to search for a place to camp and wait for a ship. They landed safely seven days later on a spot called Elephant Island.

Artist Lucia deLeiris painted this depiction of the *James Caird* sailing from Elephant Island to South Georgia Island. DeLeiris's time spent painting on the penninsula and in the Ross Sea influenced the painting's color palette and stormy sea.

Shackleton then chose five men to sail with him to Grytviken on South Georgia Island, a British-held whaling base in the South Atlantic Ocean. They fitted one of the lifeboats, the 23-foot (7 m) *James Caird*, with sails and a solid upper deck and departed on April 24, 1916. The rest of the crew stayed behind, flipping the remaining two lifeboats over to use as living quarters.

After seventeen days in cold, stormy waters, the *James Caird* arrived on South Georgia, but landed on the side opposite the whaling station. Shackleton and two others then trekked over mountains and glaciers to the other side. Shackleton asked the team to stop their trek in the morning. He knew if they had reached the whaling station, a loud whistle would soon sound to call the men to work. The team heard the blast and arrived later that day.

Shackleton still needed to rescue the men left behind on Elephant Island. But the Southern Ocean was choked with ice, and several rescue ships failed to reach the island. Finally, on August 30, 1916, the Chilean ship *Yelcho* reached the island and picked up the remaining crew. Everyone on the island was saved.

CHAPTER 2

BYRD AND THE *BEAR*

ANCHORED THIS MORNING IN THE BAY OF WHALES, DIGGING HOLES IN THE ICE WITH PICKS AND SHOVELS. THIS WAS THE ONLY WAY OF TYING THE SHIP UP ALONG THE ICE. . . . WHEN THE *BEAR* CAME UP TO THE ICE CLOSE ENOUGH FOR ME TO GET ASHORE, I WAS THE FIRST MAN ABOARD THE SHIP TO SET FOOT IN LITTLE AMERICA AND HELP TIE HER LINES DEEP INTO THE SNOW. I MET ADMIRAL BYRD; HE SHOOK MY HAND AND WELCOMED ME TO LITTLE AMERICA AND FOR BEING THE FIRST NEGRO TO SET FOOT IN LITTLE AMERICA.

—George W. Gibbs Jr., January 14, 1940

Rear Admiral Richard Evelyn Byrd Jr. is a major figure in the story of Antarctica, having led five expeditions there between 1928 and 1956. He also explored the Arctic and was reportedly the first to fly to the North Pole with pilot Floyd Bennett.

On his first expedition to Antarctica, from 1928 to 1930, Byrd and his team established a base called Little America. During this trip, Byrd, as navigator, with pilots Bernt Balchen and Harold June and photographer Ashley McKinley, became the first to fly over the South Pole in a Ford Trimotor airplane. On his second expedition,

This portrait of Rear Admiral Richard Evelyn Byrd Jr. was taken by George Gibbs Jr. on board the *Bear*.

in 1934, Byrd spent the winter by himself in a hut 100 miles (161 km) south of Little America, running a weather station. That winter he was badly sickened when carbon monoxide from a gasoline-powered generator accumulated in the hut. He recounted this experience and the other hardships of that winter in a book titled *Alone*.

As a champion of both polar regions, Byrd wrote, "There is no other music like the toneless music of millions of years of accumulated silence, through which come bars of unearthly colours. There is no need for ears to hear the fugues played on this ice organ. Here nature has set aside for man a domain of beauty and inspiration such as he cannot know elsewhere on this planet."

Byrd's adventures in Antarctica sparked worldwide interest in the continent and made him famous. In 1938 Byrd was planning a private expedition similar to his two earlier ventures when US president Franklin Delano Roosevelt (FDR) asked him to partner with a much larger US government mission. In the 1930s, the National Socialist political party, also known as the Nazis, had taken control of Germany. They were threatening their neighbors in Europe. As part of a plan to extend Germany's influence around the globe, they aimed to establish a presence in Antarctica. This motivated FDR to ensure that

Leland Curtis painted *Fogbound Shores, Antarctica* on Byrd's expedition. He also served as the artist on Operation Deep Freeze III to Antarctica in 1957.

the United States would also have a presence there. He directed the US Department of the Navy, the War Department, the Department of Interior, and the State Department to coordinate and launch the United States Antarctic Service Expedition (USASE). He asked Byrd to head the classified mission.

Byrd's first two expeditions had been privately funded and under his own direction. For this mission, the US government committed $350,000, but more funds were needed. By then Byrd was considered an American hero. He had wealthy and powerful friends. Using his connections, he was able to raise an additional $240,000 from corporations, educational organizations, and wealthy private citizens. Private donors also supplied the expedition with equipment, food, clothing, and airplanes. Byrd and FDR hoped to establish three permanent bases in Antarctica: one near the magnetic south pole, one near Palmer Land (East Base), and one near the Bay of Whales (West Base). Rotating crews would live at these bases through the summer and perhaps longer. The expedition was also tasked with exploring and mapping unknown parts of Antarctica.

Byrd insisted on taking his flagship, the *Bear of Oakland*, which he had used on his second expedition to Antarctica. After this expedition, the ship had been put on display at a museum in California. Byrd had it released from the exhibit and transferred to the US Navy. It was renamed the USS *Bear*. For this expedition, the *Bear* was fitted with a steel hull and a diesel engine, which provided a top speed of 13 knots (15 miles, or 24 km, per hour). The second ship in the expedition was the seven-year-old *North Star*, a diesel-powered, wooden ice ship that delivered supplies back and forth from Seattle to Alaska. In total, the two ships carried 125 men, 160 dogs, four airplanes, one snowmobile, a 12-ton (11 t) machine called the Snow Cruiser, and three army tanks. Fifty-nine of the men were a part of the ice parties that would live on the ice for one year. The expedition also included artist Leland Curtis, who would paint the mountain peaks, the sea, and the many shapes of the snow and ice. Curtis took two hundred canvases as well as oil paints that could withstand temperatures of –80°F (–62°C).

THE COLOR LINE

The USASE was noteworthy for many reasons, not least of which that it included the first Black people to visit Antarctica. Until then, Antarctic exploration had been almost all white. Japan had sent a team to the continent in 1911. And Dr. Louis Potaka, a New Zealand Māori, had served on Byrd's second expedition as the medical doctor. But most of those who explored Antarctica were white men from Europe, Australia, and North America.

In the United States at this time, segregation of the races was in full force. In much of the country, laws required that Black people and white people attend different schools and churches and live in different neighborhoods. White people could designate their restaurants, stores, hotels, and other places as whites-only and make it illegal for Black people to patronize those businesses. Even parks and beaches were segregated. Laws kept Black people from voting, from enrolling in some universities, and from holding jobs unless the company was Black owned. If a Black person wanted to join the navy, they had to join the Steward branch regardless of skill or interest and serve as a food preparer, food server, or custodian. Other US armed forces didn't allow Black people to participate at all.

When African American George Washington Gibbs Jr. joined the USASE, he started out as a mess attendant in the Steward branch. He and two other Black men aboard the *Bear* assisted the white officers, crew, and researchers by preparing and serving food, doing laundry, making beds, and keeping the ship clean.

Gibbs quickly learned that his mess attendant third class rank wouldn't prevent him from full participation in other assignments on the expedition. In Antarctica with Byrd, rank and race were less important than they were in the United States or on other expeditions. During his busy schedule aboard the ship, Gibbs continued to take examinations in order to get promoted.

ANCHORS AWEIGH

In the fall of 1939 George Gibbs Jr. reported to a Boston shipyard to help prepare the *Bear* for its Antarctic expedition. Gibbs chipped and scraped away old paint from the nineteenth-century ship to prepare it for a fresh paint job. At the shipyard, Gibbs met a few of his new shipmates. Harrison H. Richardson, nicknamed Jack, was the youngest expedition member at seventeen. He and Gibbs became instant friends.

George Gibbs on board the *Bear*

JIM CROW AMERICA

When George Gibbs Jr. was growing up in the United States in the early twentieth century, he lived in a world of segregation and racial discrimination. The US Supreme Court, in an 1896 case called *Plessy v. Fergeson*, had ruled that racial segregation was legal. The legal system created as a result of this case was nicknamed Jim Crow. Jim Crow separated Black and white people in almost all aspects of life. And the rules of the system were often enforced with terror. A Black person who questioned the system or failed to defer to whites might be beaten, lynched, or killed by a mob.

Under Jim Crow laws, Black people were not allowed to access or use the same spaces and services as white people. Businesses and public areas put up signs to enforce racial segregation.

In some areas of the North, racial discrimination was less extreme. But Black people and white people still lived in different neighborhoods and attended separate schools and religious institutions. A system called redlining, which prohibited the sale of homes to people of color, kept Black people out of many neighborhoods. Across the nation, Black people found that very few choices were open to them. Many institutions shut out Black people, making it hard to obtain the education to become a scientist, attorney, or other professional. Those who did manage to become professionals often had to start their own businesses or work exclusively with Black organizations. But most Black people were limited to low-paying, support positions such as factory work, house cleaning, or cargo loading.

In the mid-twentieth century, Black Americans began to fight against segregation and discrimination. They took this fight to the courts, worked to pass new laws, and protested to make changes. Other Americans, recognizing the brutality of the system, joined the cause. In 1954 the Supreme Court ruled in *Brown v. the Board of Education* that segregation of public schools was illegal. Ten years later, Congress passed the Civil Rights Act of 1964, which prohibited discrimination based on race, color, religion, sex, or national origin and made the segregation of all public services illegal. The Voting Rights Act of 1965 banned literacy tests, poll taxes, and other practices that had prevented Black people from voting. The Fair Housing Act of 1968 made redlining illegal.

These changes did not eliminate racial inequality in the United States. In the twenty-first century, many Black Americans still struggle for economic equality in schooling, jobs, and political representation. And despite changes to the law, racist attitudes and social institutions persist throughout the country. People who believe in equality continue to fight against these problems.

In November 1939, *Bear* and *North Star* departed from Boston about a week apart, heading across 11,000 miles (17,703 km) of ocean to reach the Bay of Whales. *North Star* left first. When *Bear* departed on November 22, Gibbs wrote in his diary:

> Today is very beautiful. It is snowing and there are hundreds of people waving good bye to the crew of the USS *Bear*, departing for the South Pole. Some were crying and some apparently happy. Up until today, I had wondered how one felt when people were asking for autographs and newspapers were asking for pictures. En route on the first day, caught in a windstorm.

Soon the *Bear* too was in rolling seas, which took their toll on the crew. "Everyone lost his food [vomited due to seasickness]," Gibbs wrote in his diary. "Even I lost mine."

The *North Star* sits in the Boston Navy Yard between expeditions.

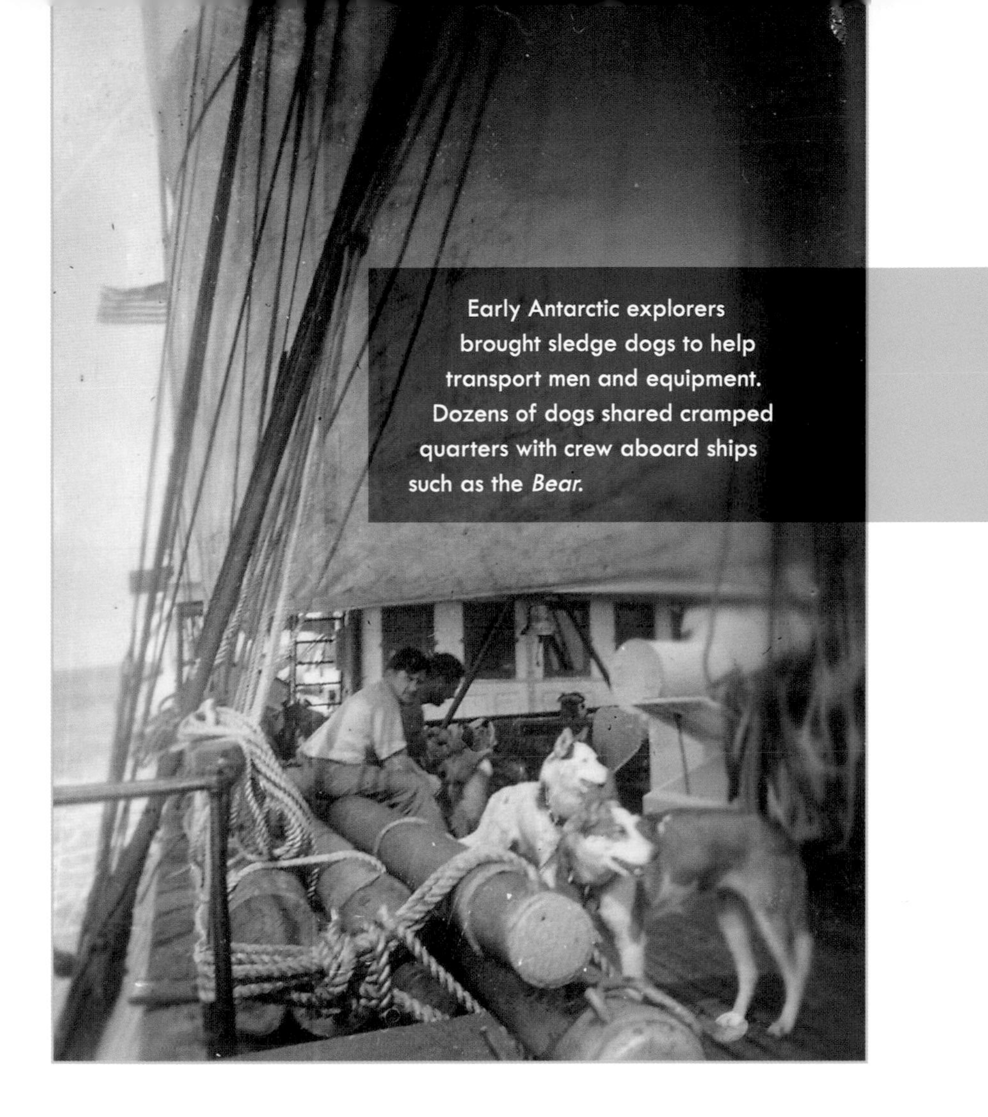

Early Antarctic explorers brought sledge dogs to help transport men and equipment. Dozens of dogs shared cramped quarters with crew aboard ships such as the *Bear.*

Quarters were tight on the *Bear*, and creature comforts were few. Gibbs recalled "taking a bath in cold water, and keeping clothes clean out of a bucket." Seventy-five huskies were housed in all parts of the ship. Gibbs wrote of the dogs "howling day and night."

The *Bear* and *North Star* made several stops on the journey south. They passed through the Panama Canal, which links the Atlantic and the Pacific Oceans, and loaded up on supplies in Chile and New Zealand. Meanwhile, the rough seas continued. "Today was another one of those rolling and rocking days," Gibbs wrote on December 3.

Researchers walk from their vessel across the ice covering the Bay of Whales, the southernmost point of open ocean on the globe.

"Water is coming all over the side of the ship. It is awfully hard for me to write at all tonight. The dogs are howling as usual."

As the *Bear* approached the Southern Ocean, the dog drivers prepared sleds and other gear for travel over the icy terrain. Each man received a pair of sunglasses and warm clothing: three pairs of long johns, two hats, and eight pairs of gloves, plus thick woolen pants and lightweight fur-and-wool parkas. Those who would be staying at the bases received the warmest clothing, including reindeer fur-lined boots.

The captain of the *Bear* told Gibbs and the others "that we would run into ice and ice bergs for Christmas." The prediction was just a few days off target. On December 28, three days after Christmas, Gibbs wrote that he had seen his first iceberg.

> Still en route to Little America. The Sea is still tossing us around like a rubber ball. We can't walk two steps without reaching for a hold on the bulkhead. I saw

> my first ice berg this morning about eight fifteen, as the ship posted within three miles [4.8 km] of it. We are doing fine, as we expect to get into Little America on January tenth. Tonight is much cooler and the wind is awfully strong.

Soon the ships had entered the Bay of Whales, an inlet of the Ross Sea. Gibbs described "icebergs and more icebergs." He wrote that the captain "was awfully busy giving right and left rudder to avoid hitting the big icebergs." About a week later, he described how "the captain climbed the [mast] as high as the crows nest to . . . direct the ship through the ice at last." He concluded, "This is the most thrilling sight I have ever experienced!" Occasionally the *Bear* even hit an ice floe, and Gibbs wrote that the sound of the crash was like thunder.

In mid-January 1940, the *Bear* reached the massive Ross Ice Shelf. The shelf towered 200 feet (61 m) tall at its highest places. It is the largest wall of ice on the planet, stretching for 370 miles (595 km) east to west.

At the edge of the ice shelf, at what would become Little America III, just north of Roosevelt Island, the *Bear* dropped anchor. The boat arrived in the middle of a blizzard, but the men were elated to finally step onto the surface of Antarctica.

CHAPTER 3

AT THE BOTTOM OF THE WORLD

EN ROUTE FOR MAGNETIC SOUTH POLE. ABOUT 1/2 OF THE "FELLOWS" WHO LIVE ON THE ICE CAME ABOARD THE *BEAR* TO HAVE COFFEE AND SAY GOODBYE TO US UNTIL NEXT YEAR AS WE ARE LEAVING LITTLE AMERICA AND WILL GO BY SHIP, AS CLOSE AS WE CAN POSSIBLY GET, TO THE MAGNETIC SOUTH POLE. THEN, ADMIRAL BYRD PLANS TO FLY OVER IT FROM THE SHIP AND EXPLORE. ALSO, HE WANTS TO OBSERVE WHY IT CAUSES THE COMPASS NEEDLEPOINT TO MOVE TEMPORARILY OFF ITS COURSE. ADMIRAL BYRD BELIEVES IN USING THE BEAR AS HIS BASE TO FLY BACK AND FORTH. WE WILL FIND THE CAUSE, THUS IMPROVING THE COMPASS MAGNETICALLY.

—George W. Gibbs Jr., February 1, 1940

EN ROUTE EAST BASE. WE CAME WITHIN TWO HUNDRED MILES [322 KM] OF THE MAGNETIC SOUTH POLE AND ADMIRAL BYRD ATTEMPTED A FLIGHT OVER THE MAGNETIC SOUTH POLE, BUT CONSEQUENTLY THE WIND CHANGES TOO OFTEN WITH THE WESTERLY GALES, AND IT IS IMPOSSIBLE WITHOUT A RUNWAY AND A TEMPORARY BASE. WE HAD TO COMPROMISE ON TIME, HOWEVER. FROM DRYGALISKIS ICE TONGUE, A BEARING WAS TAKEN ON THE MAGNETIC SOUTH POLE. . . . WE FOUND THAT IT HAD MOVED QUITE A DISTANCE SINCE THE LAST BEARING WAS TAKEN.

—George W. Gibbs Jr., February 7, 1940

Weddell Sea (off the tip of the Antarctic Peninsula) by Zaria Forman. Forman uses her pastel drawings to draw attention to the world's polar environments.

The first job for the USASE was to establish Little America III by building West Base. Earlier Antarctic expeditions had made living quarters out of wooden boards that they brought along on their ships. Sometimes they also collected local stones and even cut out ice blocks to build walls for shelters. When traveling on the ice sheet, the men slept in tents.

On the USASE, building bases was a little easier than it had been for earlier explorers. The expedition planners had put together walls, roofs, flooring, and other building parts ahead of time and loaded them onto the ships prefabricated. Before putting the buildings together on-site, the men dug about 4 feet (1.2 m) down, setting the foundations of the structures on firm ice layers beneath the snow. The floor was built 16 inches (41 cm) off the ice foundation so that warm air from the galley could circulate underneath. Walls were made of

Gibbs beside the Snow Cruiser

two wooden panels with thick insulation between them. Upon completion, the buildings were banked with snow, which helped to further insulate them from the cold. Tunnels were dug through the snow to connect them.

Gibbs would not be staying at West Base, but he helped to unload supplies and build the base. He saw the giant Snow Cruiser designed to carry a team of men across the ice. Gibbs wrote that he "rode back on a sled pulled by a tractor on the ice. It was a long cold ride, but I enjoyed it anyhow. I also drove my first sled dog team tonight." In that same diary entry, Gibbs noted, "Of course, day and night is still the same twenty-four hour daylight." It was summertime in Antarctica, when the sun never sets.

The *Bear* left a team of two dozen men at West Base. Then the ship, along with Gibbs and his fellow shipmates, continued along the coast. The ship sailed eastward and surveyed the Edward VII Peninsula, Ruppert Coast, and Hobbs Coast in late January 1940. It returned to the Ross Sea and then sailed east again to survey the unexplored coast between the Bellinghausen and Amundsen Seas.

As the voyage continued, the *Bear* went farther east and south in the Ross Sea than any ship before. Gales came and went, forcing the crew to wait for breaks in the weather. When the skies and sea were clear, Byrd's seaplane, the Barkley-Grow aircraft stowed on board the *Bear*, would take off from the water to explore inland areas.

Byrd's Barkley-Grow aircraft was ideal for Antarctic exploration. In addition to flying, the plane could be operated to use wheels, skis, or floats depending on the terrain.

Laying to where [Captain James] Cook in 1774 discovered this bay and island including an enormous ice berg and from here made his way farther south in the Antarctic than any other person so far. . . . It is our goal to explore farther south than he did. . . . This [latitude and longitude are] very beautiful with an iceberg apparently hundreds of miles long and very blue in color.

—George W. Gibbs Jr., February 23, 1940

THE SNOW CRUISER

The Snow Cruiser Byrd and his chief scientist, Thomas Poulter, designed was 55 feet (17 m) long and 15 feet (4.6 m) wide, weighed 45,000 pounds (20,411 kg), and cost $150,000 to build (about $2.6 million in twenty-first-century dollars). The vehicle was designed to house four people for a year with all their food, fuel, and equipment. Besides living quarters, it held a science lab, a photography darkroom, an engine room, and a small machine shop. Its wheels were each 10 feet (3 m) in diameter, weighed 700 pounds (318 kg), and were made from a special type of rubber that would not break or crumble in the extreme cold. Each wheel could be steered independently with its own motor, a system that would allow the Snow Cruiser to navigate difficult terrain or cross 15-foot-wide crevasses. The vehicle was designed for an average speed of 10 to 13 miles (16 to 21 km) per hour, with a maximum speed of 30 miles (48 km) per hour. Its broad upper deck was designed to carry a small airplane.

More than seventy companies pitched in to fund the building of the Snow Cruiser, which was manufactured in Chicago by the Pullman Car Company, a business that normally made railroad cars. The project went from blueprints to rollout in the span of six months. From Chicago, drivers took the newly built vehicle across the Midwest. They made test-drives on sand dunes along Lake Michigan in northern Indiana and then drove across Ohio, Pennsylvania, and New York. As it crossed the nation, roads were closed to traffic because the vehicle required both lanes. This created massive traffic jams and also drew crowds of onlookers. Finally, they reached Boston, Massachusetts, where the machine was loaded onto the *North Star.*

The Snow Cruiser and the plane that was intended to ride on the back of the vehicle when not deployed for flying.

The Snow Cruiser was supposed to drive to the South Pole and track auroras—light displays caused by charged particles from the sun hitting Earth's magnetic field—but that didn't happen. The machine turned out to be far too heavy for the Antarctic snow. It moved well over the frozen ocean ice, but on the ice sheet where Little America III was located, it sank into deep snow and couldn't climb out. After several attempts to solve its difficulties, the team at West Base simply used it as living quarters.

The expedition hoped to study the magnetic south pole, which slowly changes position as Earth's magnetic field shifts. At one point, the *Bear* sailed within 200 miles (322 km) of the magnetic pole, staying just off the coast. The seaplane took off from the water and headed toward the pole. But due to ice, dangerous winds, and sastrugi (hardened snow dunes formed by wind) near the pole, the pilot could not find a suitable landing area. Even though they could not reach the pole itself, the expedition carried equipment to measure Earth's magnetic field. Byrd's measurements showed that the magnetic south pole had moved about 30 miles (48 km) in the years since the previous measurements were made in 1934.

The *North Star* was also on the move, looking for a suitable place to set up East Base. Gibbs wrote in his diary on March 11, 1940:

> Laying to at Horse Shoe Island. We received a message informing us that the *North Star* has found a base suitable and is ready to unload, but the gales came up again about force ten. The weather here was ok this morning, and we got under way for Neny Fjord to unload also, but we decided it was best to stay here until the weather is calm enough to unload. The weather is fine this afternoon and we expect to unload the *North Star* tomorrow afternoon. Today is really beautiful.

The *North Star* and *Bear* crews teamed up to unload supplies at East Base, on Marguerite Bay on the Antarctic Peninsula, about 2,200 miles (3,541 km) from West Base. The team put together the base structures, which included a bunkhouse, a kitchen, a science workshop, a machine shop, and a storage shed. Then the two ships took off once more, leaving twenty-six men at the base.

Similar to sand dunes, sastrugi are dunes of snow shaped by wind erosion.

Once East Base was set up, both the *Bear* and *North Star* headed back to the United States. Gibbs described the scene as they said goodbye to the East Base team:

> Everyone aboard the *Bear* worked all night last night in order to leave as early as possible for the United States via a few South American ports and Panama. The ice party was sorry to see us leave, yet as we pulled out at 23 minutes and 15 seconds after 10 o'clock this morning, they waved good by and smiling said we will look for you when the next Antarctic summer sets in today. We went through a small pack ice and are at present in smooth seas. This is a beautiful night with a full moon out and every one is happy to be en route home.

A YEAR ON THE ICE

The USASE was tasked with mapping as much of the continent as possible. They mapped using ground-based surveys on the polar treks and by air using aerial photographs during their many flights. The group also performed scientific investigations to learn more about the weather of Antarctica. The expeditioners studied the continent's atmosphere, surrounding ocean, animal and plant life, and geology, as well as Earth's magnetic field. As an expedition devoted to mapping and science, it (along with the Little America I and II expeditions) set the tone for all later US visits to the continent.

Exploring and mapping new territory was the USASE's most important job. Every part of the expedition undertook this work. Some men took photographs from aircraft. Others took long treks from their bases to survey the land from the ground. Three men dogsledding from West Base were the first to set foot on Marie Byrd Land, named for Richard Byrd's wife.

INTO THE ICEBOX

What did Antarctic explorers eat? Whether aboard ship or on the polar ice, they were far from farms and cities. There were no deliveries of fresh vegetables, fruit, or milk.

On many early expeditions, explorers shot seals and penguins for food. They also ate penguin eggs. Sometimes they mixed seal meat with seal blubber to create a food called pemmican. Similar to beef jerky, it was easy to carry and provided nutrition for both people and dogs. When Ernest Shackleton's men were stranded on Elephant Island, waiting for his return, they survived on limpets (a type of mollusk) when seal and penguin meat was scarce.

The USASE crew were the first Americans to rely on frozen and dehydrated (dried) food that could be stored ahead of the voyage for many months or even years. They also ate a lot of canned foods. Frozen strawberries—thawed to room temperature—were popular on the *Bear*. The powdered eggs and milk were tolerable. Fresh milk was sorely missed.

Chocolate was another USASE staple. It is made from cacao beans, which are high in calories, fat, and nutrients. Chocolate gave polar explorers the energy they needed in cold weather. Expedition members also ate a lot of fish, caught from waters around the continent. The fish fed not only the men but also the huskies that lived with them.

Gibbs wrote in his journal about bringing this leopard seal aboard the *Bear*. Early Antarctic explorers sometimes shot and collected seals for food or research purposes.

PLAYING WITH PENGUINS

One mission of the USASE was to study Antarctic wildlife and to capture animals for US zoos. Sledge driver Hollis (Jack) Richardson made extensive notes on the birds he saw at West Base and also made sketches of their beaks and other body parts.

The most common birds at the base were Adélie penguins. They were friendly to their human visitors. The men chased them, and the Adélies chased back. When the penguins wanted to get away, they slid across the ice on their bellies and propelled themselves with their feet, a kind of movement called tobogganing. But the fun and games ended when the men captured the penguins by their flippers and tied their feet.

Expedition members would eventually take eleven living Adélies back to the National Zoo in Washington, DC. The men also captured emperor penguins for the zoo. One was an unusual golden-breasted emperor penguin they'd named Dugan. On the return trip to the US, Dugan lived in the ship's refrigerator.

Not all animals left Antarctica alive. The men at West Base sometimes shot seals, which were eaten and also fed to the dogs. They also packed up some dead seals and penguins to be stuffed and displayed in museum exhibits back in the United States. In more recent years, international law has forbidden people from moving within 6 feet (1.8 m) of Antarctic animals. The ever-playful penguins will come right up to visitors. But be warned—it's a $5,000 fine if you touch a penguin, unless you obtain a permit for research.

Adélie penguins

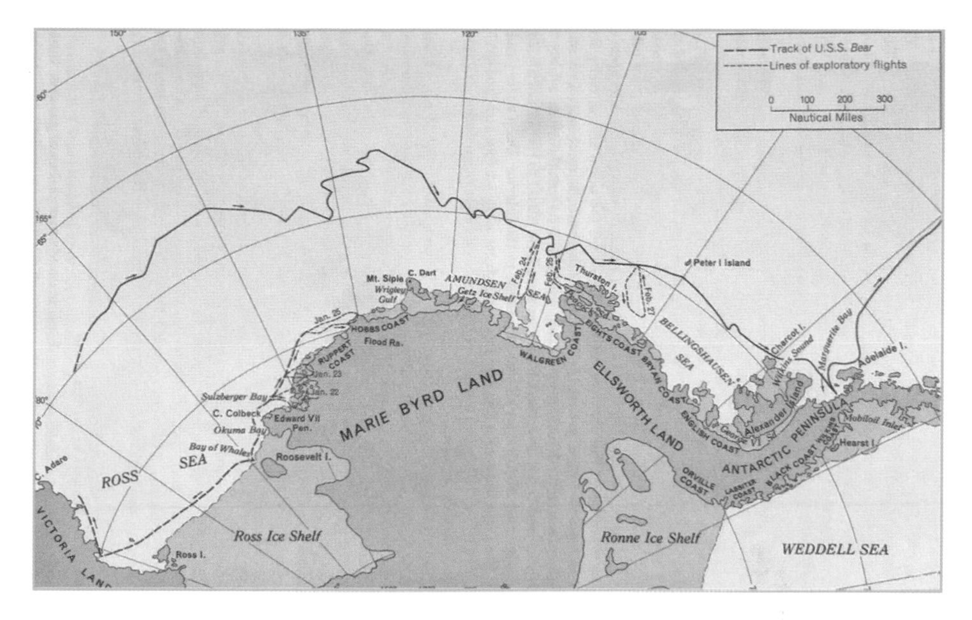

This map shows the eastward track of the USS *Bear* and the exploratory flights taken along the Antarctic coastline. As far as mapping new areas of Antarctica, the USASE was a huge success.

They made a 1,200-mile (1,931 km) round trip, the longest undertaken on the continent since Roald Amundsen's 1911 expedition. At East Base, teams of pilots and trekkers mapped 350 miles (563 km) of new coastline.

Altogether, the USASE charted more than 1,000 miles (1,610 km) of coastline and large inland regions. It identified a number of previously unknown areas, such as Alexander Island and Thurston Island. Ice parties found new access routes across the Antarctic Peninsula and mapped it using ground and aerial photos. Some scientists studied the movement of the Ross Ice Shelf. Others collected plants, animals, and rocks. Men at East Base set up a weather station. They also studied geology and did extensive research on Adélie penguins and Weddell seals. At West Base, geographer Paul Siple used the bitterly cold temperatures of Antarctica to develop the first windchill index, a measure of how wind can make cold air feel even cooler.

HOMEWARD BOUND

> All hands Aboard the *Bear*. I was up and ready for assisting in the evacuation of the East Base. Everyone was on pins and was some what worried due to the fact that the weather down here changes from a beautiful day to a horrible one is less than a half hour. However, at present the weather is fine and the entire East Base is ready to make the daring flight from East Base to Mikkelsen Island at 0530 am. There are twenty-six men and they have one airplane, which is a transport plane and will only carry 12 men with the two pilots and a limited amount of personal gear not to exceed 17,200 lbs. [7,802 km]. Including the men. They are to make two trips if they are successful.
>
> *—George W. Gibbs Jr., March 22, 1941*

The USASE was cut short by World War II (1939–1945). The war began when Germany invaded Poland in September 1939. US allies France and Britain immediately declared war on Germany. Although the expedition had been undertaken because of growing tension with Germany, by early 1941 the United States decided to recall the USASE and focus its energy on the war raging in Europe and beyond. Congress withdrew funding from the USASE, and President Roosevelt ordered the evacuation of East and West Bases. The US declared war on Japan, Germany, and Italy in the days following the Pearl Harbor attack on December 7, 1941.

Bear and *North Star*, which had returned to the United States in June 1940, traveled back to Antarctica to pick up the men. Gibbs was again a member of the crew aboard the *Bear*. He wrote that the "men who have spent a year [at West Base] look like hermits. . . . Among their first words to us on arrival at the base was "How are you boys? We are certainly glad to see you."

Evacuation at West Base went smoothly. The men efficiently skied out of the base and loaded their dogs and supplies onto the waiting ship.

Then the *Bear* sailed for East Base, and things got tricky. The sea ice surrounding Antarctica differs each year. As the ice shelf shifts and breaks apart, bays and inlets change. Areas that were once easily accessible to ships can be suddenly cut off. In Marguerite Bay, the *Bear* hit an ice pack that wouldn't move. The men at East Base knew that if the ship couldn't reach them, they'd have to spend another year in Antarctica, waiting out the darkness of winter until the ice melted the following spring.

The ship waited days in Marguerite Bay, but the ice pack still wasn't moving. The *Bear* couldn't get closer than 180 miles (290 km) from the base. After three weeks of waiting, Captain Richard Cruzen made alternative plans. They would rescue the men by airplane. Byrd was monitoring the evacuation via radio from Washington, DC. He thought the flight idea was risky. He knew that the seaplane on board the ship, the Condor, was in poor repair. But he gave the go-ahead.

The base held twenty-four men, dozens of sled dogs, equipment, personal belongings, and scientific specimens. The Condor would need to make two round trips to evacuate all the men and their gear, but there was no room for the dogs. Expedition leaders made the difficult decision to kill some of the dogs before departure to prevent them from dying of starvation. Leaders then strapped dynamite on a timer to kill the remaining dogs. They left enough time to disable the bombs in case the evacuation failed and the men had to return to the base.

Piloted by highly skilled Ashley Snow, the Condor took off for East Base. The first group of twelve men boarded the plane with their gear and made a successful flight to a landing strip on the

East Base still stands in its original location on Marguerite Bay. Aside from being used during a private 1947–1948 expedition, the base and nearly all the equipment within it were abandoned. Modern Antarctican tourists sometimes visit the base to see a time capsule of scientific research almost perfectly preserved by the cold.

Mikkelsen Islands. Snow returned to pick up the remaining twelve men and their gear. He attempted to take off with the load, but this time, the plane was too heavy. The men had to reduce the load by 500 pounds (227 kg). This meant leaving thousands of dollars' worth of goods and scientific materials on the ice.

With a lightened cargo, the plane successfully made it aloft with the second group of twelve. The two-hour plane ride lifted their spirits as they watched the spectacular scenery of ice, clouds, mountains, and sea. One last magnificent sunset appeared as if to say goodbye. To brighten spirits even more, expedition member Harry Darlington revealed a surprise for the others in his jacket. He had rescued a puppy and smuggled it onto the plane.

From the Mikkelsen Islands, the men joined the other members of their expedition, who were split between the *Bear* and *North Star*, and headed for home. Gibbs wrote happily of returning to civilization:

> At 12 o'clock today, the crew of this ship and officers were invited to a clambake, all of us aboard went (to it). We left this ship in a Chilean tug boat and were taken to a beautiful island where everyone enjoyed delicious clams and other shell fish, turkey, pork sausages, fresh sheep soup and assorted wines. There were about twenty Chilean officers and sailors with us and they gave us a wonderful time. The population is three thousand in Puerto Montt. Exports are fruit and sheep. The Chilean's are very congenial people towards any nationality.
>
> *—George W. Gibbs Jr., April 22, 1941, at the dock, Valparaiso, Chile*

Both ships arrived back in Boston in May 1941.

THE NEXT CHAPTER

When World War II ended, Antarctic exploration resumed. Richard Byrd returned to the continent again to lead Operation Highjump, a project of the US Navy. This massive expedition included forty-seven hundred men, thirteen ships, and twenty-three aircraft. On this mission, Byrd was more of a figurehead, or symbolic leader, than a commander, but his presence was still felt. Little America IV, established just north of Little America III, was one of the hubs for Operation Highjump.

Antarctic research increased tremendously with the creation of the International Geophysical Year in 1957. This twelve-nation collaboration established more than fifty scientific research stations in Antarctica, including Little America V to the east of the earlier bases in Kainan Bay (an indentation in the ice shelf edge that no longer exists). Since the late 1960s, the United States has operated three permanent

ALL HANDS ON DECK

When Gibbs joined the USASE, he was delighted to find that there was no segregation between Black and white people on the USS *Bear*. Although Gibbs was an officer's attendant, he pitched in with other responsibilities, such as unloading equipment, building landing strips for airplanes, and catching penguins. He appreciated the opportunity to contribute and be respected for his work.

Occasionally, though, he was harassed by officers, who belittled him by calling him "boy" and "darkie," offensive terms for Black people. Captain Cruzen, who commanded the *Bear* along with Byrd, intervened on Gibbs's behalf. He reprimanded one officer who had insulted Gibbs: "Officer, We don't talk like that here. On this ship, we all help out when needed. Regardless whether you're in the galley [kitchen], making the officers beds, or adjusting sails, if an extra pair of hands is needed, it doesn't matter what color the hands are. Do you understand?"

Cruzen also singled out Gibbs to receive an award for meritorious service on the voyage in acknowledgment of his hard work, energy, and loyalty to his shipmates. Most of the team cheered when he earned his award. Later in the expedition, he received the honor a second time.

Antarctic stations: the Amundsen-Scott South Pole Station, McMurdo Station on Ross Island, and Palmer Station on an island near the Antarctic Peninsula.

THE CONTINENT CALLS

Antarctica continued to attract scientists and adventurers, who arrived wanting to test their courage and strength in the polar wilderness. In 1989 American adventurer Will Steger led a seven-month dogsled

Ann Bancroft and her all-women team reached the South Pole in January 1993. Bancroft continues to embark on expeditions and speaks on issues related to the environment and LGBTQ equality.

traverse of Antarctica. His team of six explorers traveled 3,741 miles (6,021 km) with no assistance from mechanized vehicles. In 1993 American Ann Bancroft led an all-female ski expedition to the geographic South Pole. (Several years earlier, she had traveled by foot and by sled on an expedition to the North Pole.) In 1997 Norwegian Borge Ousland traveled 1,864 miles (3,000 km) across Antarctica from sea to sea. On this trip, he hauled his sledge behind him in the fashion of Robert Scott and his men. In 2001 Bancroft returned to Antarctica with Norwegian Liv Arnesen. This time, they skied all the way across the continent.

The twenty-first century has seen additional adventurers tackle the continent in various ways: with sea kayaks and sailboats, on foot and by bicycle, and with kite-assisted skis. Some have traveled alone, some in teams. Some have retraced the paths of early explorers. Others have tried to set records as the youngest or the fastest traveler to make the journey. As these adventurers show, the unknown southern land has not lost its allure.

GEORGE GIBBS: GROUNDBREAKER

> En route Valparaiso S.A. Today we are well on our way and with 2/3 [of our trip] behind us. We expect to arrive on Sunday April 21. Tonight finds us still with good weather although she is still rolling. Oh well, this is the *Bear*. There are two Naval officers aboard U.S.N. who at times make this cruise very hard for me and if it wasn't for the Captain aboard here, I would certainly be put ashore on arrival or at least another U.S.N ship. But for his sake and mother's, I will try to stick it out, at least until I get home.
>
> —George W. Gibbs Jr., April 18, 1940

> At Valparaiso Chile. Was up this morning early to get the ship in shape for visitors and do my daily routine. This morning at eleven we were anchored off from the dock about two hundred feet [61 m] and I went ashore in a taxi. Bought a watch and went to Vina Del Mare, which is eight miles [13 km] from Valparaiso. Went sight-seeing and had dinner at the Hotel Higgins where the Admiral [Byrd] is staying. In the U.S. it would have been impossible for me to be as welcomed [as I felt there].
>
> —George W. Gibbs Jr., April 23, 1940

George Washington Gibbs Jr. was born in Jacksonville, Florida, in 1916. As a teenager during the Great Depression (1929–1939), he decided to leave school to join the Civilian Conservation Corp (CCC) to earn money for his family. There he did the hard labor of building bridges, planting trees, and clearing forests. He joined the US Navy after one year.

In 1939 twenty-two-year-old Gibbs reenlisted in the US Navy after his first four years of service. He applied to be part of Byrd's

Gibbs on the *Bear*

United States Antarctic Service Expedition and became one of the forty navy men accepted out of two thousand applicants to join the crew of the USS *Bear*. The experience was the highlight of Gibbs's life. It shaped his attitude toward those who discriminate against people because of race. By 1941 he had learned there were people outside of the United States who treated him as an equal. This realization made him commit his life to serving the human race.

Following his Antarctic journey, Gibbs remained in the navy. He finished high school aboard a ship with Brooklyn Tech's GED program. He fought in World War II aboard the aircraft carrier USS *Atlanta*. During the Third Battle of Savo in 1942, the *Atlanta* was attacked by forty-nine Japanese shells and one torpedo. Gibbs handed out life jackets to his shipmates as the ship sunk into the Coral Sea. He and other survivors spent a night and a day in shark-infested waters before being rescued by the US Marines. For two years, he fought with the marines in foxholes and on a patrol torpedo boat until he could get back to the United States.

Gibbs remained in the navy after the war and became a chief petty officer. In the late 1950s, he applied to join Byrd V, or Operation Deep Freeze. But this time, he didn't get the assignment. He decided to retire from the navy after twenty-four years of service and rejoin civilian life. He earned his college degree from the University of Minnesota; moved to Rochester, Minnesota; and became a human resources manager at IBM, a technology company. After eighteen years there, Gibbs opened his own employment company, Technical Career Placement. It was his lifelong dream to become a business owner, and the business continues to serve its community.

Gibbs was an active leader in numerous community organizations including Toastmasters, the YMCA, Kiwanis Club, the University of

Minnesota Alumni Association, the American Red Cross, and the United Methodist Church. In the 1960s, he joined the movement for civil rights and cofounded the Rochester branch of the National Association for the Advancement of Colored People (NAACP). In 1974 the local Elks Club, a community service organization, used blackballing (a secret ballot process in which as few as three people can block membership) to refuse membership to Gibbs because of his race. Gibbs protested this discrimination, and the story made national news. He received hundreds of calls of support from the Minnesota community, including the club leadership, who invited him to join. Due to his activism, the group opened its doors to Black people that same year.

George W. Gibbs Jr. died in 2000. His humanitarian legacy lives on. George W. Gibbs Jr. Elementary, Minnesota's first Gold LEED green-building certified school, is named after him. So is Rochester's beautiful Gibbs Drive. One of his highest honors is a place now named Gibbs Point. This rocky point, located on the northeast side of Horseshoe Island and Marguerite Bay along the Antarctic Peninsula, is named for Gibbs—the first Black man to set foot on Antarctica.

CHAPTER 4
SECRETS OF THE ICE

LAYING TO AT HORSE SHOE ISLAND. THIS MORNING THE *BEAR* MADE AN ATTEMPT TO ESTABLISH AGAIN THE EAST BASE. BUT DUE TO GALES AND ICE WE WERE FORCED TO TURN BACK TO OUR ANCHOR, AFTER GETTING UNDER WAY EARLY THIS MORNING. TONIGHT WE ARE ANCHORED, THE WIND IS HOWLING TERRIBLY AND THIS IS THE DARKEST NIGHT I HAVE EVER SEEN IN THE ANTARCTIC, WHICH MEANS THAT IT WON'T BE LONG BEFORE IT IS DARK FOR TWENTY FOUR HOURS FOR ABOUT FOUR MONTHS.

—George W. Gibbs Jr., March 9, 1940

Modern scientific expeditions have taught us much about Antarctica. Scientists can now trace the story of the continent back to its beginning. Geologists believe that more than two hundred million years ago, all the land on Earth was part of a supercontinent called Pangaea. The northern region of Pangaea, called Laurisia, contained lands that would later split apart to become North America, Europe, and most of Asia. The southern part of Pangaea, called Gondwanaland, combined landmasses that would later become

South America, Africa, India, Australia, and Antarctica. A large body of water, the Tethys Sea, was wedged between Laurasia and Gondwanaland. The last vestige of this waterway, which once stretched from Europe to Tibet, is the Mediterranean Sea.

Scientists think that about two hundred million years ago, Pangaea began to split apart. Laurasia separated from Gondwanaland, and eventually these two landmasses broke up into the continents we know today. How do scientists know that the continents split apart? One clue comes from simply looking at a map of the world. The shapes of coastlines suggest that continents now separated by oceans probably once fit together. For example, the outline of eastern South America and western Africa are a very close match. This fact was noted by geographers for centuries before the theories of continental drift and plate tectonics were developed. A similar fit exists for the Wilkes Land coast of Antarctica and the southern coast of Australia.

Other evidence for Gondwanaland takes the form of fossils—the remains of ancient plants and animals. Fossils show us that life-forms on all the southern continents were at one time similar. This suggests that these areas were once all part of the same landmass. The fossils also show that the climates in these areas were once very different from what they are now. Fossils of tropical plants and animals found in Antarctica were one of the early supporting clues to plate tectonics and the true age of Earth: 4.6 billion years, not thousands or millions of years as in early estimates.

Even data from satellites orbiting Earth point to the existence of Gondwanaland and even more ancient continental arrangements. One satellite, the Gravity Field and Steady-State Ocean Circulation Explorer (GOCE), mapped the gravitational field near the South Pole and across East Antarctica in great detail. The gravity data revealed the remains of several ancient tectonic plates beneath East Antarctica. These plates, large movable sections of Earth's crust, are left over

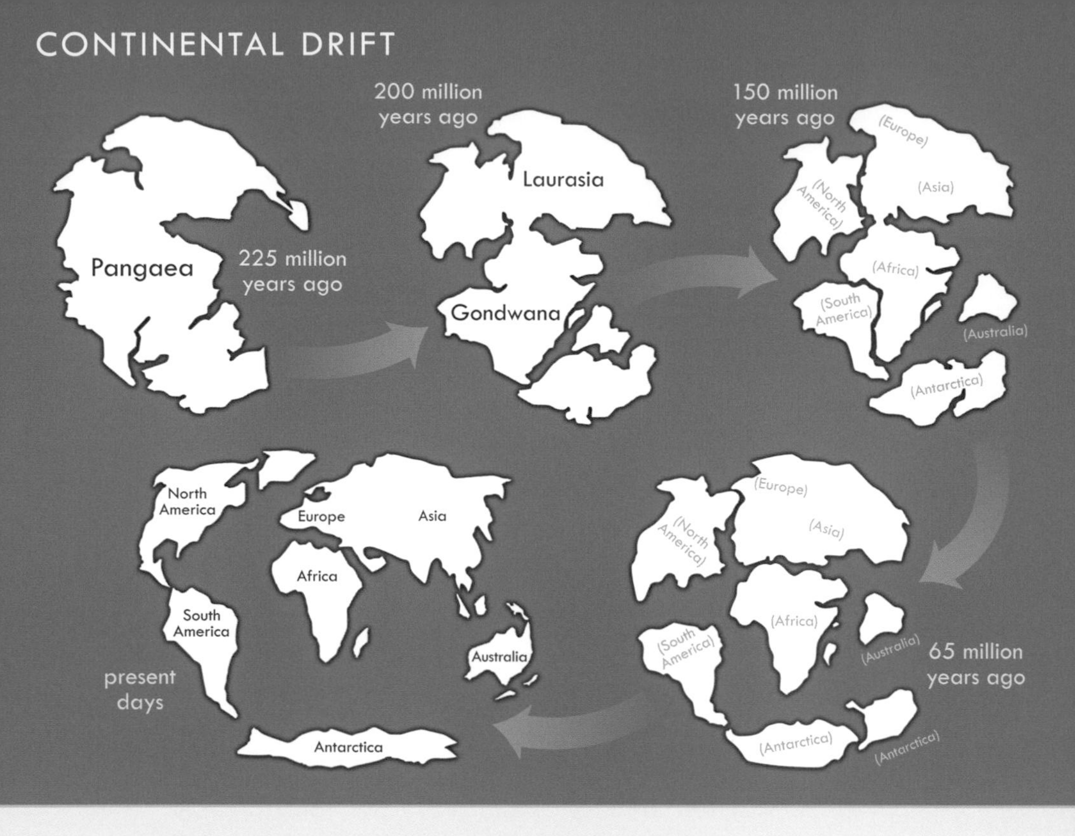

This diagram shows how the ancient continent Pangaea gradually split up into the seven continents of the current geological era.

from a time before Australia, India, and other continents were joined together as Gondwanaland—an even earlier version of the puzzle of continental pieces.

WARM FRONT

Earth's climate has changed many times since the planet formed. Our planet has gone through numerous cycles of warming and cooling. Modern Antarctica is too cold for most plants and animals, but Earth was much warmer at the time when the southern continents were linked together as Gondwanaland. The higher temperatures made Gondwanaland a suitable home for many trees and plants. Dinosaurs and other animals roamed on a completely ice-free land.

The evidence for this ancient life, again, comes from fossils.

Antarctic explorers and modern researchers have found dinosaur bones and other fossils that prove that the continent was once home to many plants, birds, and other animals that thrived in a much warmer climate. Robert Scott gathered fossils on his 1911 Antarctic expedition. They included the fossilized remains of a tree resembling modern-day beech trees (*Nothofagus antarctica*) from just a few million years ago. In 2011 Chilean researchers found a fossilized, football-sized egg on an island off the coast of Antarctica. Dating back sixty-eight million years, it might be the egg of a mosasaur, a large ancient sea reptile.

DEEP FREEZE

Beginning about forty million years ago, Earth entered one of its cooling periods. This was a result of the atmosphere losing carbon dioxide as the Himalayan mountains began to form. Glaciers began to cover some of the mountainous areas in the interior of Antarctica, now buried under the ice sheet, but the climate was still warm enough for many plants and animals. Over time Earth cooled further, and the Antarctic ice sheet began to form. By three million years ago, ice covered nearly the entire continent.

Several factors make Antarctica the coldest place on Earth today. Because it sits atop a pole, the amount of solar energy reaching the continent's surface is smaller than in the tropics, the areas closer to the equator. In winter, when Antarctica is tilted away from the sun, much of the continent sits in total darkness for several months. Without the sun's rays, and with very dry conditions, Antarctica becomes extremely cold. In addition, a band of water called the Antarctic Circumpolar Current (ACC) flows through the ocean around the continent, traveling from west to east. The current creates a barrier around Antarctica, keeping more tropical waters from the north from reaching the continent. The surrounding ocean surface freezes in winter, creating a fringe of floating sea ice that extends for hundreds to thousands of miles from Antarctica's edge.

Only a few types of living things can survive in modern Antarctica's vast inland areas. These include simple organisms such as algae, lichens, and bacteria. Birds sometimes visit the interior, even reaching the South Pole, but many need food from the coast or areas farther north to survive. Penguins spend most of their time along the Antarctic coast, and many other birds such as terns and petrels spend summers in Antarctica. The ocean and coastlines around the continent are suitable for larger, more complex living things, including fish, whales, seals, and a host of creatures living on the seafloor.

LIE OF THE LAND

Ice covers about 98 percent of the Antarctica. If all the ice were to melt, it would be easy to see the various geographical features beneath it. These include all the features other continents have: mountains, plains, canyons, valleys, and thousands of islands lining the coasts. Antarctica even has liquid-water lakes and rivers beneath the ice.

Geographers separate the continent into eastern and western sections based on the hemisphere the area lies upon. East Antarctica makes up about two-thirds of the continent. The area's ice sheet holds three-quarters of all the ice on Earth. Much of the bedrock in East Antarctica is above sea level, which means that its ice sheet is more stable and less susceptible to rapid melting due to a warmer ocean. Along its coastline are dozens of mountain ranges with spectacular jagged cliffs. Its interior ice plateau is a vast, almost featureless white expanse, including Wilkes Land (named for nineteenth-century US naval commander Charles Wilkes) and Queen Maud Land (named for a Norwegian queen). The Transantarctic Mountains divide East and West Antarctica. This range hugs the coast of the Ross Sea and Ross Ice Shelf on one side of the South Pole and continues along the Ronne Ice Shelf and the Weddell Sea coast on the other side. The interior of West Antarctica is Marie Byrd Land. Other features of West Antarctica include the Antarctic Peninsula, the Siple Coast, Pine Island

In the summer, ice covering the open ocean around Antarctica melts, leaving chunks of ice bobbing in the water. Because of this, the continent is much easier to reach in the summer.

Bay, Ellsworth Mountains, and thousands of large and small islands. West Antarctica's bedrock, however, mostly lies well below sea level. It is thought to be an ancient seabed. Though the bedrock sits below sea level, it is buried in ice that reaches more than a mile (1.6 km) above.

From the days of the first explorers, Antarctica has presented a variety of amazing and unusual landscapes. The Ross Ice Shelf is one of the most famous. On his 1840 expedition, James Clark Ross named the shelf the Barrier because its ice cliff creates a wall at the southern edge of the Ross Sea, preventing ships from traveling farther south. However, the Ross Ice Shelf offers the shortest route to the South Pole if a path onto the top of the ice can be found. To reach the pole, explorers in the early twentieth century had to leave their ships at the edge of the ice shelf and travel over the ice on foot or by dogsled. The ice shelf is moving and changing all the time as it flows northward. Gigantic plates of ice break off the front every decade or so, and the bending, grinding, and cracking of ice flowing can be heard for miles. Antarctic adventurers describe the sound as the roar of a cannon or the screams of a monster.

Another landscape formed by the unique conditions found only in Antarctica are the so-called Dry Valleys. Discovered in 1903 by the first Scott expedition, the Dry Valleys are windswept areas on the southwest coast of the Ross Sea where extremely dry conditions have evaporated nearly all the glacier ice, leaving a rocky surface with many unusual features. In the Dry Valleys, the terrain is changed mainly by wind, not water. The Dry Valleys near McMurdo Station and several similar wind-swept regions, such as the Vestfold Hills and the Bunger Oasis near East Antarctica's northern coast, have had no rain for centuries. These areas are bitterly cold and dusty, similar in some ways to the surface of the planet Mars. They are some of the most extreme environments

Due to dry and cold conditions, parts of Wright Valley have remained largely free of ice and snow for thousands of years.

Erebus is one of the few consistently active volcanoes in the world. Most volcanoes lie still for long periods of time and erupt only occasionally. Erebus bubbles, spits gas, and hurls blocks of red-hot rock almost continuously.

on Earth, and only the hardiest plant and primitive animal life-forms manage to survive. Despite the dry conditions, trickles of glacial meltwater form lakes during the very warmest weeks of the summer. The extremely salty lakes stay liquid but are capped by thick layers of ice that remain in place year-round. The lakes support an unusual ecosystem of thick algae mats and diatoms. In the Dry Valleys' soil there are lichen, bacteria, and a few flatworm creatures called nematodes.

HIGHS AND LOWS

Antarctic mountains are particularly dramatic because they are shaped by extreme glacial erosion. In total, Antarctica has more than one hundred mountain ranges and many isolated mountain peaks (called "nunataks," a word from the Inuit language). Mount Erebus on Ross Island is one of the continent's most famous mountains. It measures 12,448 feet (3,794 m) high and is the southernmost active volcano on Earth. Ernest Shackleton and his team of men were the first to climb Erebus during the 1907 *Nimrod* expedition.

Modern scientists have located more than one hundred volcanoes on the continent, many of them along the coast or in the interior of West Antarctica. These mountains embody a dramatic combination of fire and ice. When a volcano erupts, hot gases and steam surge upward.

But in the cold Antarctic air, the steam quickly condenses and freezes to form icy towers called fumaroles. Active volcanoes also produce lava. In most volcanoes, the lava is trapped inside an inner chamber, which is covered by cool, solid rock. But Mount Erebus's lava chamber is open at the top, so scientists can peer down inside to study it. The lava lake on top of Erebus is thought to be several miles deep, with a temperature of 1,700°F (927°C).

During the International Geophysical Year from July 1957 to December 1958, scientists from the Soviet Union discovered a mountain range that is completely buried within the ice sheet—the Gamburtsev Mountains of East Antarctica. Because they are entirely below the surface, no one has actually seen them. Before this discovery, scientists thought that most of the landscape below the ice of the Antarctic interior was flat. But the Soviet team, using explosive charges and seismometers that listen through the ice for echoes from the land underneath, found a jagged landscape that reached upward thousands of feet. The ice around the Gamburtsev Mountains is 2 to 3 miles (3.2 to 4.8 km) thick, but the peaks climb to within half a mile (0.8 km) of the ice surface. The range is as big as the Pyrenees mountain range that forms the border between France and Spain. In 2008 an international team of scientists led by US geologist Robin Bell conducted a major airborne mapping campaign of the area. They found that the mountains were reshaped by small alpine glaciers tens of millions of years ago, and that being encased by the ice sheet has preserved the landscape for far longer than the life span of other mountain ranges on Earth. Other studies have predicted that the ice near the mountains, in the bottom layers of the ice sheet, may be the oldest ice on the continent. This ice may have fallen as snow as much as 1.5 million years ago. It could contain traces of dust and air that can tell us about Earth at that time.

Several other vast mountain ranges rise above the Antarctican ice. Mount Vinson, part of the Ellsworth Mountains in West Antarctica, is the highest peak on the continent, measuring 16,860 feet (5,139 m) above sea level. Another mountain in the Ellsworth range is shaped like a four-sided pyramid. The mountain is so evenly shaped on all four sides that some people say that aliens or an ancient human society must have built it. But geologists provide a less fantastical explanation. They say that millions of years of wind and weather carved the mountain's geometric shape. Mountains in other places on Earth also take the shape of pyramids, with two or three uniform sides. But this mountain, so far unclimbed and unnamed, is the only one known to have four equal faces.

Antarctica also has deep canyons, valleys, and even impact craters. East Antarctica's Wilkes Subglacial Basin is one of the largest craters on Earth. It measures about 300 miles (483 km) wide and over 500 miles (805 km) deep. Two Dutch explorers discovered the basin on a trip across Victoria Land in 1959. Scientists think that a large meteorite created the crater when it crashed into Earth about 250 million

The ice-covered mountains of the Antarctic Peninsula are only about 620 miles (1,000 km) away from the southern tip of South America.

years ago. To gouge out such a giant basin, the meteorite must have been about 30 miles (48 km) wide. In addition to gouging out the crater, scientists think that the force of the impact filled the skies with dust and debris. This material would have blocked out the sun, causing land plants and animals to die. The meteorite might have caused one of Earth's mass extinction events, when many life-forms suddenly die out all at once.

Another large canyon—more than twice as long as the Grand Canyon in Arizona—is completely buried under the ice in Princess Elizabeth Land in East Antarctica. As with the Gamburtsev Mountains, explorers walking across this region would likely think they were crossing flat terrain. But satellite images tell a different story. Scientists use these images to look for subtle features in the Antarctic ice sheet that indicate what the landscape is like below the surface. They look for clues, such as flat spots or wavy areas, that might reveal buried lakes or mountains. In the case of Princess Elizabeth Land, satellite images from a number of sensors, as well as airborne radar profiles of the ice sheet, led scientists to map the ice-covered canyon as well as additional smaller canyons and lakes beneath the ice. Radar profiles, similarly to seismic profiles, are made by sending strong radio pulses into the ice and using antennae to listen for echo returns from layers in the ice or the ground below it.

WATER BENEATH THE ICE

Scientists have found that Antarctica has about four hundred subglacial (under-ice) lakes. The idea of unfrozen lakes thousands of feet below the ice might seem strange. Why doesn't the water in the lakes freeze? Several factors keep it liquid. For one thing, heat from deep inside Earth keeps the water above the freezing point. Pressure from the weight of the ice above also changes the temperature at which water freezes. At the bottom of the ice, the freezing point of water can be two to three degrees lower than its normal freezing temperature. In some cases, as in the Dry Valleys, the subglacial lakes are very salty. Salt water has a lower freezing point than fresh water, so that even at the temperature where fresh water would normally freeze (32°F, or 0°C), salt water stays liquid.

Many of Antarctica's subglacial lakes are connected by subglacial streams. Scientists have been able to find out more about the lakes and rivers by drilling through the ice. Russian scientists have studied subglacial Lake Vostok near the South Pole. Close to the size of Lake Erie in North America, but much deeper, it's the fourth-largest lake on Earth in terms of volume (the amount of water it holds). Covered by 13,000 feet (3,962 m) of ice, Lake Vostok is 143 miles (230 km) long and 31 miles (50 km) wide. It has a maximum depth of 2,625 feet (800 m). The lake has been sealed over by ice for more than fifteen million years. Another lake, Lake Mercer in West Antarctica, is covered by 3,000 feet (914 m) of ice. Drilling into Lake Mercer in 2018, scientists found the remains of tiny land crustaceans (tardigrades) that may have been washed off of nearby mountains in a past climate when the ice sheet was much smaller.

DEEP AND DANGEROUS

Crevasses are fissures in the ice that come in all sizes, from narrow cracks you can jump across to expanses as large as canyons. The polar regions and other icy parts of the world are home to these fractures, which open due to movement of the ice. Crevasses make some areas of Antarctica dangerous to explore. They can be difficult to see and detect, especially when the wind forms snow bridges over their openings.

Early Antarctic explorers relied on their sled dogs to discover crevasses to avoid falling into them. In the twenty-first century, scientists use radar, satellite images, and other technology to find crevasses. Using portable radar systems, modern-day explorers can detect crevasses up to 160 feet (49 m) below the ice.

The *Bear* and its crew encountered many crevasses on their expedition. Though dangerous, crevasses give researchers clues about the movement of ice in an area.

In more than one hundred years of Antarctic exploration, hundreds of people, dogs, ponies, skis, tractors, and sledges have fallen into crevasses. Many people have fallen to their deaths. George Gibbs fell into a crevasse in 1940 when he first arrived on the continent. On January 18, he wrote in his diary:

> Anchored in Bay of Whales. The wind is calm today and of course it is very cold, but a beautiful day in Little America. As far as one can see is ice covered with snow, hills and high mountains of ice ridges. It is 12 o'clock midnight, the sun is very high and the wind is terrific. I walked more than two miles [3.2 km] on the ice tonight, and in doing so, I went down to my waist in a crevasse, but came out ok.

MAPPING ANTARCTICA

Ancient mapmakers imagined that Earth held a southern continent. They placed it on their maps although no one knew for sure whether the land existed. But in the nineteenth century, explorers such as James Weddell, Fabian Gottlieb von Bellingshausen, and James Ross confirmed that indeed a continent was there. They began mapping the coastline of Antarctica using their own observations and measurements. Inland expeditions filled in the map even further. When Byrd and other pilots flew over Antarctica, they filled in more blank areas by taking aerial photographs. But even as late at 1971, a map of Antarctica was mostly featureless and white—because that is what pilots saw and what aerial photographs could show.

In the late twentieth century, satellite images revolutionized the work of mapmakers. Radar maps and digital satellite images compiled of the entire continent began to reveal the structure of the ice sheet at a level of detail not seen before. Previously unknown mountains, canyons, and lakes beneath the ice sheet were revealed by subtle topography on the surface. In the first decade of this century, new maps from Landsat and other satellites used sophisticated processing to show Antarctica in exact true color in digital images with excellent detail. In the fall of 2018, scientists at the University of Minnesota and Ohio State University showed the world the most detailed map of Antarctica made to date: the Reference Elevation Model of Antarctica, or REMA. Created by combining nearly 190,000 satellite stereo images and processed to show the continent in 3D, the map shows surface features that are no bigger than a car. It reveals icy ripples, snow formations, melting glaciers, and cracking ice shelves. The map helps scientists better track the melting and flowing ice and other changes in Antarctica caused by rising global temperatures.

A year later, back again at Little America, Gibbs wrote on January 10, 1941, that "the accumulated ice has been cracking which caused many crevasses and one of our men fell in a crevasse twenty-five feet [7.6 m], but we got him out unharmed, except for a few scratches."

Others had more harrowing experiences. In 1947 American Harries-Clichy Peterson, part of Finn Ronne's expedition to Antarctica, fell about 130 feet (40 m) into a crevasse and was stuck there upside down for nearly twelve hours. Finally, a search party found him and hauled him to safety with ropes. Others weren't so lucky, and even in recent years, crevasses have claimed the lives of explorers in Antarctica. In 2016 Gordon Hamilton, a climate scientist and glaciologist, died when returning by snowmobile to camp after surveying part of the Ross Ice Shelf. In 2017 David Wood, a Canadian helicopter pilot, landed to refuel his aircraft, fell into a crevasse, and died from his injuries.

WATER AND SKY

Much of the magnificence of Antarctica is the water in all its forms, from the ice-encased coastlines to the ice- and snow-covered interior. The ocean around Antarctica is filled with ice, from massive ice shelfs to towering icebergs to small ice floes. In the ocean near the coast of the continent, the frozen ice on the sea provides a different kind of seascape, as changeable as the clouds overhead.

Sea ice is seawater that has frozen solid. As the ocean begins to freeze, the water becomes filled with tiny feather- and needle-shaped bits of ice, called frazil ice. As this begins to collect and thicken on the surface, one of two things can happen. If there is wind or waves, the crystals begin to bunch together in circular patches called pancakes or silver-dollar ice or even lotus leaf. If the air and water are very still, the crystals freeze into thin plates like glass, called nilas. Eventually these are pushed and welded together into larger pieces. A small piece of sea ice is an ice floe. Pack ice refers to areas where many ice floes are bunched together.

Different types of ice form depending on the temperature, wind, and water conditions of a given area. Pancake ice (*top*) is a kind of new ice, or ice that forms over a short period of time. Nilas ice (*bottom*) can form into stronger ice crusts over time.

Ice shelves are thick slabs of glacier ice that jut out from the coast. They form where glaciers flow out over cold water to cover bays, seas, or even large areas of open ocean. Ice shelves can be up to 1,500 feet (460 m) thick. Icebergs are huge chunks of floating ice that have broken off of glaciers or ice shelves in a process called calving. Some are like large islands, tens of miles across. Many of Antarctica's icebergs are tabular, or flat-topped. They form when large top-to-bottom crevasses, called rifts, split the shelf ice apart and a flat sheet of ice hundreds of feet thick drifts away from the fissure. Icebergs contain far more ice than meets the eye. The portion you see floating above the water might be less than one-seventh of the size of the overall berg. The rest of it lies underwater.

The skies above Antarctica are just as spectacular as the water surrounding it. During full winter darkness, the winter-over staff at the bases (those who live in Antarctica during the winter months to maintain buildings and equipment) can often see the aurora australis, or southern lights. (The equivalent at the North Pole is the aurora

Some icebergs around Antarctica have been frozen for so long that the ice forms distinct layers, looking somewhat like sedimentary rocks. Scientists can extract information about the climate conditions of the past by analyzing the ice layers.

The northern and southern lights can occur at any time of day, throughout the year. However, they are best viewed by humans during the winter and late at night when the sky is darkest. Visitors can witness the bright greens and purples of the aurora australis, the South Pole's equivalent of the northern lights, or aurora borealis.

borealis, or northern lights.) These beautifully colored, glowing areas in the sky appear when charged particles (electrons and protons) from the sun hit gases in Earth's uppermost atmosphere. When this happens, the gases ionize and emit light, making the skies light up with green, red, and purple streaks. The lights move and change shape under the influence of Earth's magnetic field. They can stretch across the skies for thousands of miles.

In daylight, visitors can see extraordinary clouds in the skies above Antarctica. Noctilucent clouds form when ice crystals in the air mix with dust from meteorites. *Noctilucent* means "night-glowing." These dramatic, icy-blue clouds form high in Earth's atmosphere. You can only see them during the polar twilight, when the sun is sitting just below the horizon. This happens before Antarctica goes into full darkness in winter or as it emerges into round-the-clock sunlight in summer.

Nacreous clouds are sometimes called "mother-of-pearl" clouds because of their iridescent colors.

Nacreous clouds look like fire in the sky. *Nacreous* means "pearly" or "iridescent." They too form in the cold air high above Earth, and they appear at the end of winter. Another dazzling display occurs when sunlight bounces off sea ice and lights up the undersides of clouds. This phenomenon is called iceblink.

POLE PARTY

When people think of Antarctica, they often think of the geographic South Pole. The pole is one of the most inaccessible places on Earth. The land beneath the ice at the South Pole sits only a few hundred feet above sea level, but it's covered by more than 9,000 feet (2,743 m) of ice. This thick layer of ice makes the South Pole extremely cold, since air gets colder the higher you go in the atmosphere. How cold is cold? The lowest temperature ever recorded at the South Pole was –117°F (–83°C). The warmest ever was 9.9°F (–12°C).

The magnetic south pole is always on the move and is actually quite far from the geographic South Pole. Scientists find the magnetic pole using a device called a dipping needle, which is similar to a compass.

A HANDY MAP

Want to see a map of Antarctica? You can make your own using your left hand. Climate scientist Ted Scambos came up with this easy trick. Start by holding your left hand palm up in front of your face. Make a loose fist, folding your fingertips into the middle of your palm. Stretch out your thumb so it's pointing toward the left. In this position, your thumb represents the Antarctic Peninsula. Your knuckles stand for Queen Maud Land in East Antarctica. The flesh below your thumb is West Antarctica, and the right side of your palm is Wilkes Land in East Antarctica. The area above your wrist, between your thumb and your lower palm, is the Ross Sea. The Weddell Sea is in the crease between your thumb and index finger.

Copy the diagram above using your own hand as a shorthand for remembering the geography of Antarctica.

The magnetic pole is the point on Earth's surface where the dipping needle, tracking the magnetic lines of force, points directly downward. The pole has moved more than 900 miles (1,448 km) since members of Ernest Shackleton's team first tried to locate it in 1909. At that time, the pole was in Victoria Land, a part of the continent south of New Zealand. By 2020 it was in the Southern Ocean, off the Claire Coast in Wilkes Land, and moving in the direction of western Australia.

LOOKING BACKWARD

The ice sheet covering Antarctica has been building up for about forty million years. But the ice sheet doesn't contain only ice. Trapped in the thick icy layers are dust, minerals, fossils, air bubbles, gases, and microbes that at one time were on or near the surface of Antarctica. This buried material is like the contents of a time capsule. It can tell us what conditions were like in Antarctica hundreds of thousands of years ago.

To uncover that information, scientists drill into the ice sheet and remove large cylinders of ice called ice cores. Then they study the ice layers, from the deepest (oldest) to the uppermost (most recent), to see how Antarctica's climate, air, and water have changed since the ice sheet first formed. The layers of ice give clues to the planet's warming and cooling trends, recent ice ages, and the modern climate. The dust and other minerals tell scientists about the composition of the air and water around the continent many thousands of years ago. Scientists also drill into the ice to capture and analyze air.

Drilling a deep ice core involves heavy machinery. At Lake Vostok and other bases, scientists use giant hollow drills to cut a core from the ice. Additional powerful machinery is needed to pull cores up to the surface. Then the cores are shipped to laboratories around the world for study. They must be kept frozen at temperatures far below freezing to ensure that the ice and its contents do not change before being analyzed.

Lynn Montgomery, an affiliate researcher with the University of Colorado, holds an ice core from the Antarctic Peninsula. It is important to drill deep in order to capture many layers—and ages—of ice.

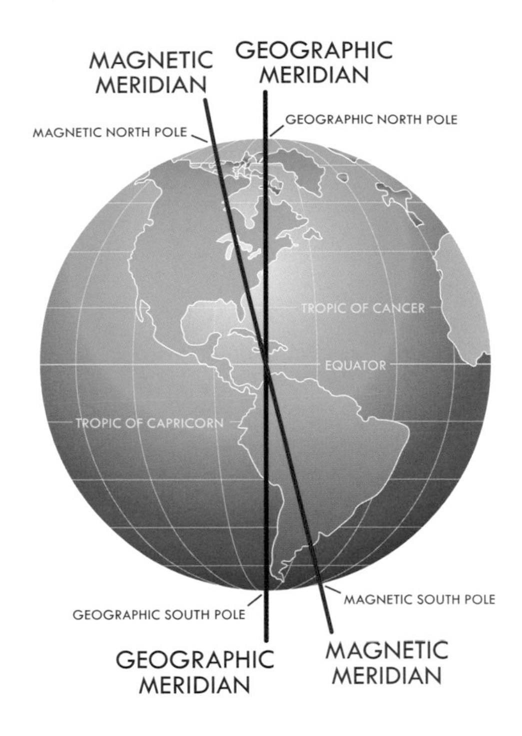

The locations of the geographic poles differ from those of the magnetic poles by about ten degrees. The magnetic poles shift due to the movement of molten iron in Earth's outer core.

Explorers have focused their energy on reaching the geographic and magnetic poles, but scientists have identified other poles as well. The southern pole of inaccessibility is in the deep interior of the continent at the point most distant from any point on the coastline. The instantaneous south pole is the exact point where Earth's axis (an imaginary line through the planet, around which it spins) meets the surface. Since Earth wobbles a little as it turns on its axis, the instantaneous south pole moves too. It makes a counterclockwise circle, about 60 feet (18 m) across, mirroring the course of Earth's wobbly spin. This path is called the Chandler Circle. The south pole of balance sits in the center of the Chandler Circle. Finally, scientists track the geomagnetic south pole. This is the south pole of Earth's overall magnetic field—a field created by currents in the liquid iron core in Earth's interior. This magnetic field is not perfectly symmetrical, so the location of the magnetic pole, where the lines of force are vertical, is not in the same place as the center axis of the average, overall magnetic field of Earth.

CHAPTER 5

FLORA AND FAUNA

LAYING TO AT HORSE SHOE BAY. TAYLOR WAS SENT OVER TO THE NORTH STAR TO HELP UNLOAD IT, LEAVING THE STEWARD AND I TO DO THE WORK ABOARD OUR SHIP. AFTER SECURING THE PANTRY AT ONE O'CLOCK P.M., I WENT ON A SIGHT SEEING TRIP TO GO TO HELL ISLAND. WE WENT IN A WHALEBOAT WITH A SMALL MOTOR ATTACHED TO IT. THE MOTOR STOPPED SEVERAL TIMES BEFORE WE REACHED THE ISLAND AND WE HAD TO PUSH ICE OUT OF OUR WAY TO REACH THE ISLAND. IT THE ISLAND HAS A MOUNTAIN ATTACHED TO IT AND MANY ODD SHAPED ICEBERGS AND MOUNTAINS, WHICH ARE VERY BEAUTIFUL ON THE ISLAND. WE FOUND HUNDREDS OF PENGUINS AND ROCKS SHAPED LIKE DIAMONDS AND OTHER ODD SHAPES.

—George W. Gibbs Jr., March 15, 1940

If you were plopped down in the interior of Antarctica, you might think that nothing could ever live there. The land is covered with ice as far as the eye can see. Temperatures lower than –50°F (–46°C) are common. How can anything survive?

Thousands of species do live in Antarctica. Some of them are unlike the living things found anywhere else on Earth. They are specially adapted to survive in the bitterly cold climate. From bacteria to lichens, algae, seaweed, mosses, birds, and fish, more creatures survive in the world's harshest environment than one would expect. Most of these organisms live in the continent's ice-free zones.

The smallest living things in Antarctica are viruses and bacteria. These microscopic organisms live everywhere on Earth, in the hottest places, the driest places, and even deep under the ice of Antarctica. Scientists have drilled through more than 2 miles (3.2 km) of ice to Lake Vostok and other subglacial Antarctic lakes and pulled samples of lake water up to the surface. They found the water was teeming with bacteria. Scientists have identified four thousand species of microbes in one subglacial lake.

Lichen and algae are other simple life-forms that make their home in Antarctica. Lichens are found growing on rocks on the small portion

In summer months, Antarctic algae adapted for growing on snow transform the color of the ice depending on the species and conditions.

An electron microscope image of a tardigrade

of Antarctica that's not covered by ice—mostly along the coast and in the mountain ranges. Algae also grow on snow and ice. In some places the snow in Antarctica looks red or green from algae growing there. The most common land plant on Antarctica is moss, which grows along with lichens among the rocks in coastal areas. Antarctica is also home to two species of flowering grasses in the northernmost parts of the peninsula and its adjacent islands. The continent has one native insect, the Antarctic midge, which lives in pools of meltwater around penguin and seal feces.

Tardigrades, sometimes called water bears, are some of the most interesting animals in Antarctica. These microscopic creatures have eight legs, claws, and chunky bodies. Found all over the world, they are practically indestructible. Scientists label them extremophiles because they can survive in extreme heat, extreme cold, and other harsh conditions. They live in environments that would kill almost every other plant or animal, such as places filled with poisonous chemicals or deadly radiation. Tardigrades have even been sent into outer space and have come back alive. In Antarctica, tardigrades live on the rocky coasts and in lakes. They feed on algae, bacteria, and plant cells. In laboratory tests, tardigrades have survived after being frozen to temperatures as low as –458°F (–272°C), just above the lowest temperature possible (absolute zero, a temperature at which there is no thermal energy whatsoever). Even the coldest Antarctic conditions don't pose a problem for them.

BIRD-WATCHING

In October and November, part of the Antarctic spring, a wide variety of birds begin to breed on the continent's coastlines. They include shags, skuas, terns, snow petrels, and albatrosses. A shag is a kind of cormorant that lives on Antarctica's islands and the Antarctic Peninsula. Wandering albatrosses are the largest seabirds on Earth, with wingspans up to 12 feet (3.7 m) wide. They feed on fish, squids, and tiny sea creatures called krill and can fly up to 6,200 miles (9,978 km) at a stretch.

Skuas, a type of seagull, are excellent flyers and have brownish-gray bodies and wings. They are true scavengers, often raiding Adélie penguin nests for eggs and small chicks, and can venture well into the interior of the continent. They are even found as far south as the South Pole. Skuas also follow ships at sea. The crew aboard the *Bear* often saw as many as fifty at one time. The birds hovered nearly motionless in the wind while the men were cutting up seal meat for their dogs. They waited to feed on the discarded pieces of the seal carcasses.

At 4 inches (10 cm) long, the snow petrel is the smallest Antarctic bird. It spends July and August in the Arctic and then heads south, arriving in Antarctica in December. In February it heads north again, completing one of the longest migratory paths of any animal. Pure white, with black eyes and blue-black-colored feet, snow petrels eat whale, penguin, and seal carcasses. They breed on remote mountain cliffs, some of them just a few hundred miles from the South Pole.

When people think of Antarctica, they often think of penguins, although these birds live in many other parts of the Southern

A shag sits on its nest near a group of Adélie penguins.

Hemisphere too. Penguins are found on the islands and coasts of New Zealand, Australia, South America, Africa, and even the Galápagos Islands near the equator. Penguins can't fly, and they waddle when they walk, but they are agile swimmers. Five species of penguins live in Antarctica. Adélie penguins can be identified by the white rings surrounding their eyes and their relatively simple black-and-white plumage. Chinstrap penguins are easy to identify because of the dark bands beneath their beaks that look like permanent smiles. Macaroni penguins have orange feathers that stick up like eyebrows on top of their heads. Emperor and king penguins look quite similar. Both are larger, standing up to 3 feet (0.9 m) tall and weighing as much as 84 pounds (38 kg), and both have yellow feathers around the neck.

Adélies are the most common penguins in Antarctica. They build nests of pebbles along the coast and were well known in the camps of early Antarctic explorers. Emperors are the largest penguin species. Male emperors spend the winter in large groups, standing close together in crowds called huddles. As they huddle, they shuffle so that every penguin spends time in the warm inner areas of the huddle. The males tend to the eggs laid by the females through much of the winter and early spring as they wait for the females to return with food. The emperor parents recognize their chick's call for food among fifty thousand birds. Like all penguins, emperors feed on fish and other sea creatures. They can dive 1,800 feet (550 m) below the surface of the water—the deepest of any birds—to catch fish and other prey. Adélies and emperors are the only penguin species that live exclusively on the mainland of the continent.

SWIMMERS

Sea life is abundant in the waters of Antarctica. It includes snails, sea urchins, jellyfish, and much more. All these organisms rely on others in their environment for food. At the bottom of the food chain are phytoplankton, tiny organisms that get their energy from sunlight and minerals in the water. They provide food for krill, shrimplike animals that cluster in large schools. Krill are small, only 1 inch (2.5 cm) long, and usually pinkish. They are eaten by larger sea creatures, including whales, seals, penguins, and fish.

The largest animal in the world can be found in the oceans surrounding Antarctica. The captivating, powerful, and endangered blue whale can measure up to 100 feet (30 m) long and can weigh 170 tons (154 t). One blue whale can eat 8,000 pounds (3,629 kg) of krill in a day. Early Antarctic explorers commonly saw blue whales on their voyages. USASE members watched them circling their ships, mating, and feeding. The *North Star* even crashed into a blue whale.

Orcas, or killer whales, eat up to 300 pounds (136 kg) of food per day. They mainly eat penguins, fish, seals, and the smaller species of whale. Even though they rarely attack humans in the wild, Antarctic explorers keep them at a safe distance. Humpback whales are curious and affectionate toward humans. They have a diverse diet of fish and krill. When food is scarce, they can live off the fat stored in their bodies.

Leopard seals eat penguins and can be eaten by whales and sharks, although this is rare. The crew from the *North Star* captured a leopard seal during the USASE trip. The seal fought to defend itself, but the crew managed to snare it with a rope and drag it back to their ship behind a sled dog team. The seal ended up in the National Zoo in Washington, DC.

STRANGE CREATURES

Some species of Antarctica are unlike animals found anywhere else on Earth. For instance, the blackfin icefish has clear blood, transparent bones, and no scales to protect its body. Scientists are studying this strange fish to learn more about how it survives. They think the fish might shed light on human blood and bone diseases such as anemia and osteoporosis.

Another fascinating fish is the Antarctic toothfish (renamed Chilean sea bass in restaurants), a large and somewhat fearsome-looking fish that lives near the seafloor in southern South America and all along the Antarctic coastline. It can grow up to 8 feet (2.4 m) long and can weigh hundreds of pounds. Its blood contains a kind of antifreeze to help it survive the frigid temperatures of the Southern Ocean. Antarctic octopus species have a similar survival system. A substance in their blood called hemocyanin keeps them from freezing in the icy seas.

Other weird Antarctic creatures include giant sea spiders—some as big as dinner plates—and glass sponges (*right*), whose skeletons contain the same substance that makes up glass. All these organisms have evolved over millions of years to survive in the dark and icy waters of Antarctica.

flower-basket, like all sea sponges, captures nutrients from water flowing through its skeletal body.

Crabeaters are the most abundant seals on Earth. They number about fifty million, and most of them live in Antarctica. Despite their name, they eat mainly krill, not crabs. A crabeater weighs about 660 pounds (299 kg) and grows to be 8.5 feet (2.6 m) long.

Weddell seals are great swimmers. They can dive 2,000 feet (610 m) below the water and ice and remain there for forty-five minutes. This ability helps them hide from orcas, leopard seals, and other predators. Fish are their preferred prey, although they also eat crustaceans, octopuses, and other sea life.

Elephant seals are the largest seals of all, and the deepest divers. Adult males can weigh up to 8,000 pounds (3,629 kg) and measure 14 to 19 feet (4.3 to 5.8 m) long. They can be identified by their stubby, floppy snouts that look like shortened elephant trunks; females have shorter snouts and are about half as large. Elephant seals can dive to tremendous depths, as deep as 5,000 feet (1,524 m), and are fast swimmers. They eat penguins, fish, krill, and other seals.

Elephant seals spend the vast majority of their lives underwater, being capable of holding their breath for over one hundred minutes. They come ashore to rest and breed.

CHAPTER 6

HUMAN-MADE TROUBLE

THIS MORNING THE WIND DIED DOWN AND WE GOT UNDER WAY TO ESTABLISH AN EAST BASE WHICH WAS FOUND LAST P.M. AT NENY FJORD. TODAY ADMIRAL BYRD DECIDED TO MAKE THE EAST BASE THERE WITH THE NORTH STAR AS A TEMPORARY BASE, UNTIL THEY GET SETTLED ON THE MOUNTAIN VALLEY. WE ARE ANCHORED HERE, STANDING BY, IN CASE OF A DESTROYING WIND BLOWING IN THE BASE DIRECTION. THERE IS LOTS OF ICE AND ICEBERGS IN THIS AREA AND IT GETS COLD ENOUGH EVERY NIGHT TO FREEZE THE ICE, MEANING THAT IF WE DON'T GET OUT SOON WE WILL GET FROZEN IN UNTIL NEXT YEAR.

—George W. Gibbs Jr., March 7, 1940

TODAY WE ARE STILL ANCHORED AT HORSE SHOE ISLAND AND THE NORTH STAR WITH MOST OF THE ICE PARTY WHO IS TO ESTABLISH THE EAST BASE. THERE WAS AN ATTEMPT TO ESTABLISH THE BASE AT NENY FJORD, BUT FOUND UNSUITABLE BY ADMIRAL BYRD AND THE BASE LEADER MR. BLACK. THIS INVOLVED THE BARKER GROVE IN ANOTHER FLIGHT TO FIND A SUITABLE BASE ELSEWHERE. AT LAST WE HAVE FOUND ONE, WE THINK, AFTER ABOUT A THREE-HOUR FLIGHT. IT IS ESSENTIAL TO UNLOAD AT ONCE AND CLEAR OUT, OR WE WILL GET FROZEN IN.

—George W. Gibbs Jr., March 8, 1940

Earth has entered a new, human-caused warming period. Whereas warming and cooling cycles were natural phenomena earlier in Earth's history, this newest warming period results from a different kind of change in the atmosphere. In the two centuries since the beginning of the Industrial Revolution, advanced commercial agriculture and farming, and the age of technology, we have made an impact on many natural processes on Earth. How we do this is clear. When we burn fossil fuels—coal, oil, and natural gas—extra carbon dioxide enters the atmosphere. Carbon dioxide is a greenhouse gas, which means that it traps heat that is trying to leave Earth, holding it in the atmosphere and slowly warming the planet. Increased emissions of methane, in part from leaking oil and gas wells and from factory farm animals, also contribute to the greenhouse effect. The gases lead to irregular warming and cooling patterns of the atmosphere, stratosphere, and ocean; changing wind patterns; lower humidity and soil moisture; and changes in ocean circulation.

Although ice coverage in Antarctica changes with the seasons, climate change has also diminished the amount of ice over time and caused huge sheets of ice to break apart as never before.

The extra heat trapped by carbon dioxide and methane is having a big impact on the climate all over Earth. In some places, rising temperatures are leading to more powerful storms or heavier rainfall. In other places, the changing climate is causing more frequent droughts—long periods with little or no rainfall—or more intense wildfires. In the Arctic and Antarctica, glaciers and ice shelves are melting more rapidly, and in many places this leads to the glaciers flowing faster and dumping more ice into the ocean. As polar ice melts into the ocean, sea levels rise. So far, this rise is just a few millimeters (less than an eighth of an inch) each year, but the amount is rising decade by decade. Many low-lying islands and coastlines have started to flood with higher high tides (so-called king tides) and storm surges.

Warmer conditions in the Pacific and tropical areas are causing the winds surrounding Antarctica to blow faster than before (the so-called westerly winds, a belt of windy conditions over the Southern Ocean). These stronger winds change ocean currents on the surface, and as a result, deeper water is moving closer to the icy coasts in some areas. This deep water is several degrees above freezing, and when it reaches the coast of Antarctica, it melts the underside of glaciers that touch the ocean. This is especially true in Pine Island Bay and the Thwaites Glacier region. In a few other areas, such as the northern Antarctic Peninsula, warmer winds and melting on the surface create summer melt ponds on the ice. The Antarctic Peninsula has 674 glaciers, and 90 percent of them are showing some summer melting and retreat. Because water is denser than ice, the water on the surface tends to flow down into crevasses, and can further fracture the ice.

GLACIERS: GOING, GOING . . . GONE?

Antarctica holds some of the largest glaciers in the world: Pine Island and Thwaites, both situated on the coast of West Antarctica. Pine Island Glacier is more than 1 mile (1.6 km) thick and covers an area of about 68,000 square miles (176,119 sq. km). The nearby Thwaites

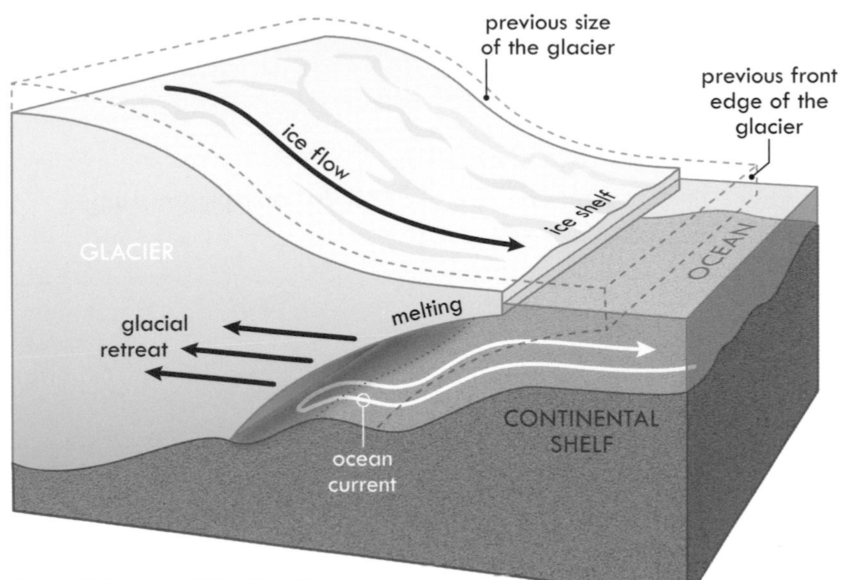

GLACIAL RETREAT

Where the glacier ice meets the ocean, it thins and begins to float. This creates a thick, floating plate of ice called an ice shelf. In some areas of Antarctica, winds have changed ocean currents so that warmer deep water now reaches the place where the ice begins to float. The water, a few degrees warmer than freezing, melts the ice, removing the lower part of the glacier and causing the ice above to flow faster.

Glacier is even larger. Each glacier covers an area similar to the US state of Florida or the island of Great Britain. But because of climate change, both glaciers are getting smaller.

Glaciers are not fixed in one place. They move slowly across the land, flowing downhill like thick syrup or wet cement. Pine Island and Thwaites flow from the center of West Antarctica's ice sheet out toward the Amundsen Sea. When the glaciers reach the sea, the ice keeps flowing outward, extending like an icy tongue over the water. Some of the ice melts when it reaches the ocean. At the same time, fresh snow falling on Antarctica piles up on top of the inland portions of the glaciers. In previous decades, the inland snowfall balanced out the amount of ice melting and falling into the sea, so the glaciers stayed about the same size year after year.

But as Earth's climate has warmed, the oceans have gotten warmer too, and the ocean currents have changed. While the surface of the ocean surrounding Antarctica still freezes every winter, and the upper layer of water is still very cold, the deeper ocean layers near the continent are warmer than before. So when Pine Island and Thwaites now meet the Amundsen Sea, the deep water beneath them is warmer than it was in the past. The warmer water melts more ice from the bottom of the glaciers as they reach the ocean. The glaciers flow more rapidly as a result, drawing more ice from within the West Antarctic Ice Sheet. This melting is now taking place much faster than new snowfall is building up the higher areas of the glaciers. So the glaciers are shrinking.

Scientists now study the melting glaciers using satellite images and other satellite and airborne technology. They use airborne radar and seismic echo sounding to measure the thickness of the ice and the shape of the bedrock below. The data shows that Pine Island and Thwaites are sending more ice into the sea than any other glaciers in the world, and they are susceptible to even faster ice flow in the future.

In 2018 a team of one hundred British and American scientists went to Thwaites to further examine the melting. The scientists measured air and water temperature and mapped the bedrock beneath the glacier. They drilled through as much as 2,000 feet (610 m) of ice to place scientific instruments inside the glacier and in the ocean beneath it. They also studied the surrounding ocean using automated submersibles, or remotely controlled undersea vehicles. The scientific studies revealed that the glacier was losing more ice than ever before. In the 1980s, the glacier lost about 4 billion tons (3.6 billion t) of ice per year. By 2019 Thwaites was losing close to 50 billion tons (45.4 billion t) per year.

The shrinking of the glaciers is distressing because the melting ice released into the oceans leads to rising sea levels. If Thwaites were to melt completely, sea levels would rise by 1.6 to 2 feet (0.5 to 0.6 m). This would spell catastrophe for coastal cities and low-lying islands.

Coastal flooding has become more common as ocean temperatures rise.

For example, parts of Miami, Florida, are impossible to protect from increasingly high tides because the soil there is so deep and porous that walls and dikes cannot block the rising water. In southern Louisiana wetlands, some people have been forced to leave their land, becoming some of the first climate refugees. The Biloxi-Chitimacha-Choctaw Native Americans lived on their island in the Gulf of Mexico for centuries. In 1955 it was 22,400 acres (9,065 ha), nearly twice the size of Manhattan. Now only 320 acres (130 ha) remain.

People in many other parts of the world have seen their homes and farms swallowed by rising seas. Many citizens of Pacific and Indian Ocean–island countries have moved to islands with higher ground. In the coming years, major world cities such as Venice, Italy; Shanghai, China; Jakarta, Indonesia; and San Jose, California, are likely to experience severe flooding as sea levels rise. Vast areas of these and other cities might become uninhabitable, and city planners are already thinking about the cost of rebuilding sea defenses, port areas, and airports.

Engineers and scientists are motivated to find solutions to these tough problems. Some engineers have even proposed building a giant submarine wall in the ocean in front of Thwaites Glacier, 50 to 60 miles (80 to 97 km) long and 1,970 feet (600 m) deep, to hold back the warmer ocean water that is melting the glacier. Most scientists say the solution would be too costly and would not last long. And Thwaites Glacier is just one of many large glaciers that need to be protected. Scientists say that the only sure way to slow the melting of glaciers is to slow climate change. Slowing climate change means curbing our use of fossil fuels, reducing carbon dioxide and methane emissions, and continuing to create new solutions for conservation and energy production.

REPERCUSSIONS

Melting glaciers are only one of the changes brought to Antarctica by rising temperatures. As the continent warms, its permafrost areas, or regions of permanently frozen soil, are also thawing and changing. And as carbon dioxide levels rise in the atmosphere, ocean waters are absorbing more carbon dioxide from the air. The extra carbon dioxide in the water has changed the chemical balance of the Southern Ocean, and the world's oceans in general, making them more acidic. Some sea life can't survive in the changing ocean environment. Warmer temperatures and more acidic waters are threatening the health of corals, sponges, krill, crabs, scallops, and other small creatures. If these animals die, the animals that rely on them for food, including whales and many other creatures, will die as well.

These changes from warming temperatures, melting ice, thawing soil, and acidifying ocean have altered Antarctica's ecosystems, the communities of plants, animals, water, and land that work together as a unified whole. For example, some of Antarctica's animals have begun to migrate in response to warmer temperatures. Adélie penguins, which thrive on ice-covered coastlines, have started to move south as their past icy habitats (original territories) have melted away. Gentoo penguins, which prefer open water and rocky shorelines, are also moving south from their original habitats into areas where the ice has melted. As animals move from their original habitats, they disrupt the ecosystem of the new area and that of their old habitat.

PEOPLE = PROBLEMS

In addition to climate change, people have created other problems for Antarctica. For centuries, fishing, whaling, and sealing crews have hunted the animals of Antarctica. In many areas, sea life has been greatly depleted. Some species, including seals and several types of whales, have been hunted to near extinction. Many are still endangered or vulnerable.

Over more than one hundred years of Antarctic expeditions, explorers and scientists have left food and human waste at their base or in the sea. As tourism to Antarctica increased in recent decades, the problems worsened. Their ships sometimes leaked oil and other pollutants into the sea. Even modern-day researchers often use fossil fuels to power machinery and heat buildings or tents in Antarctica and, in the past, used dangerous chemicals to run some experiments and to drill through the ice. The substances don't break down easily in the freezing temperatures of Antarctica. They remain there, frozen or otherwise stable, on the pristine landscape. Early researchers left old refrigerators, fueling equipment, synthetic clothing, and flame-retardant fabrics on the polar ice. Eventually, wind, sunlight, and water eat away at these materials, releasing toxic chemicals as they break down.

A new effort to clean up the areas around the polar bases began in the late 1990s and continues today. Large piles of aging barrels, old machinery, and other waste were removed from the older bases like McMurdo Station. New methods of managing waste were instituted, and field activities were made less damaging. Now many field researchers use solar power to charge batteries and to run small machinery. Most Antarctic bases have extensive recycling systems, with up to 80 percent of the recyclable materials collected and shipped north. Several new bases are powered completely by clean energy.

However, pollution has also reached the continent from other parts of the globe. Polluted air and water know no boundaries. They travel from crowded cities to remote, empty places, including Antarctica. For instance, in the mid-twentieth century, farmers in the United States and elsewhere used a chemical called dichlorodiphenyltrichloroethane (DDT) to kill insects that damaged their crops. But scientists learned that DDT was also harmful to birds and fish. The chemical would become concentrated in the animals' bodies and negatively impact reproduction. So the United States and many other countries banned the use of DDT in the 1970s. But even though the chemical was outlawed and was never used in Antarctica, it ended up in the bodies of penguins living there. How did that happen? DDT got into the food chain. Birds and fish ate insects that had been poisoned by DDT. When the birds and fish were eaten by larger animals, the DDT got into their bodies as well. All the world's oceans and air are connected, meaning that DDT 3.2 entered the food chain of Antarctica even when it was used far away. And though DDT has been banned (except for small amounts in some countries for malaria control), other pesticides (substances used to kill insects) are still used. These too have been found in the bodies of penguins, whales, and other life in Antarctica. The chemicals can make animals sick and damage their reproductive systems, preventing them from having offspring.

Plastics have also found their way to remote Antarctica. People use tons of plastic, and they don't always recycle it or dispose of it properly. Because plastic takes so long to break down, it ends up in fields and rivers. It blows in the wind and gets washed into the oceans. There it bobs in the water and eventually breaks down into tiny pieces called

About 8 million tons (7 million t) of plastic ends up on the ocean every year. By 2050, unless trends change, there will be more plastic than fish in the ocean by weight.

microplastic. Since the oceans are all connected, waves and currents carry microplastics all over the world. Even in far-off Antarctica, microplastics have entered the food chain. Fish and other sea creatures unknowingly take in plastic bits when they eat. Even small amounts of plastic can sicken and kill the animals that eat it.

A HOLE IN THE SKY

British adventurer Robert Swan was the first person to walk to both the South Pole and the North Pole. He made the South Pole walk in 1987, pulling a sled behind him and tracing the footsteps of Robert Scott. During the South Pole walk, Swan spent a long time outdoors during the Antarctic spring, when a seasonal hole opens in Earth's ozone layer above Antarctica. Harmful ultraviolet (UV) rays from the sun traveled through the hole and damaged Swan's skin and eyes.

The ozone layer is a region of the stratosphere many miles above Earth's surface that contains traces of the gas ozone, a molecule made of oxygen. The ozone layer normally protects plant and animal life from dangerous UV rays, because ozone absorbs UV light. UV radiation can cause skin cancer and other illnesses in humans and can harm and kill plants. Without the ozone layer, many of the animals and plants that live on Earth's surface would die.

In the 1980s, scientists discovered that the ozone layer was thinning. Human-made chemicals called chlorofluorocarbons (CFCs) were depleting the ozone by causing chemical reactions that broke it down. In the air above the polar regions, CFCs even made brief holes in the ozone layer beginning in the spring seasons of the late 1970s. Holes in the ozone layer above the Arctic are relatively rare, occurring only once every few years. But in the Antarctic, a hole in the ozone layer has occurred every year since at least 1979. At the time, CFCs were widely used in refrigerants, air conditioners, aerosol spray cans, plastic packaging, and other materials. After scientists figured out how CFCs were damaging the ozone layer, countries worldwide signed a treaty that banned the chemicals from most uses. The ozone layer has thickened slightly since CFCs were banned, but some of these chemicals remain in the atmosphere for decades or even centuries, and the ozone layer has not fully recovered. The ozone hole over Antarctica continues to reappear between August and October each year. In recent years, some of the springtime ozone holes have been smaller than the largest ones in the 1990s and early 2000s.

CARING FOR THE CONTINENT

Antarctica is not a nation with a government and elected officials. Instead, nations elsewhere on Earth have agreed to cooperatively govern the continent. This collaboration began with the International Geophysical Year (IGY), which actually lasted a year and a half, from July 1, 1957, to December 31, 1958. During the IGY, twelve countries—Argentina, Australia, Belgium, Chile, France, Japan, New Zealand, Norway, South Africa, the Soviet Union, the United Kingdom, and the United States—set up more than fifty scientific

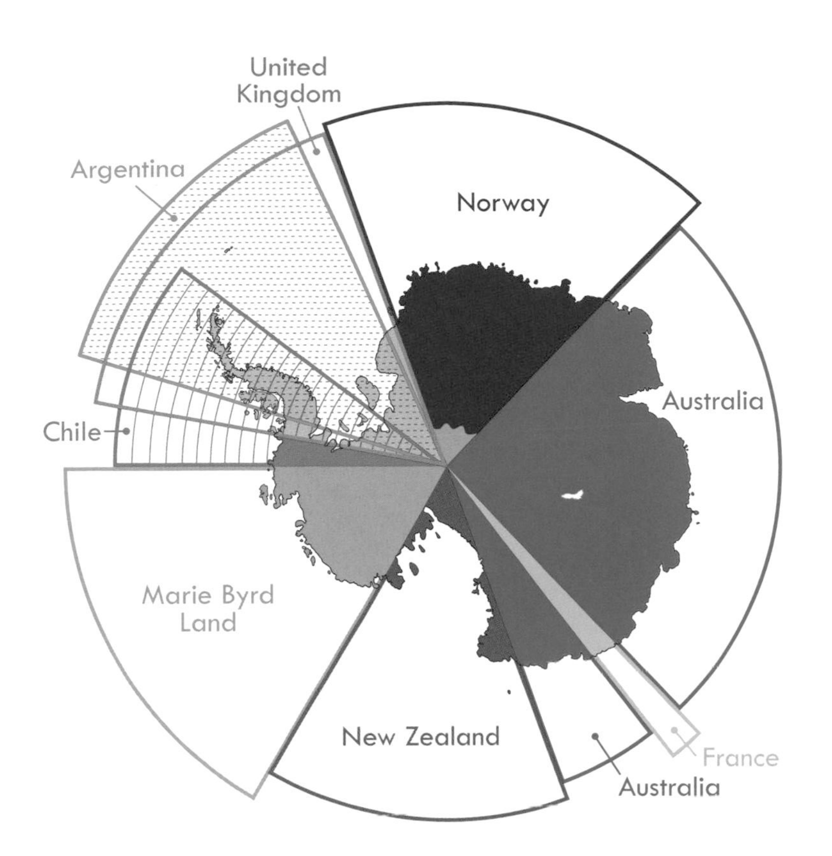

Different countries have laid claim to sections of Antarctica. These nations meet each year to discuss scientific research, environmental protections, and other issues related to Antarctica.

bases on Antarctica. Seven of the twelve countries had existing claims on sections of Antarctica, shaped like slices of a pie. These territory claims were based on past exploration of the coast, the southern extension of their borders, or on areas where past fishing or whaling activities had occurred.

In 1959, after the IGY ended, the twelve countries signed an agreement called the Antarctica Treaty. The treaty prohibits nations from claiming sections of Antarctica as their own territory, and it suspends the earlier territory claims indefinitely. It states that people can use the continent only for peaceful purposes such as scientific research. It forbids military forces on Antarctica unless they are part of scientific expeditions. It also outlaws the testing of nuclear weapons there, and it forbids the disposal of dangerous nuclear waste on the continent. The treaty requires scientists working in Antarctica to share the results of their research, so that all people can benefit.

Representatives from the twelve nations signed the treaty on December 1, 1959. It took effect on June 23, 1961. In the following years, additional countries signed onto the treaty. By 2020 more than fifty nations had signed. The treaty has been praised as a model of international cooperation for the peaceful use and conservation of a major global resource.

The Antarctic Treaty System (ATS) was an important first step in caring for the southernmost continent. But in the late twentieth century, it became clear that more needed to be done to protect the continent's land, waters, and living things. Other human activities, such as oil spills from ships and overfishing, were harming Antarctic wildlife.

To address these and other problems, in 1991 the Antarctic Treaty nations signed an addition to the treaty called the Protocol on Environmental Protection. The protocol designates Antarctica as a natural preserve and prohibits mining there for fifty years. It states that all activities, including tourism, must be carried out with a minimal impact on the land. It restricts the harming of wildlife on the continent. It calls for the cleanup and removal of abandoned old buildings and waste dumps. It prohibits the dumping of waste from ships in Antarctic waters and calls for the immediate cleanup of any oils spills. The protocol also limits human access to certain parts of the continent, especially its interior. The ATS comes up for renewal in 2048. Antarctica continues to need protection for the benefit of the entire world. Will we recommit to the treaty? Will humanity continue to protect the continent's flora and fauna, and to use Antarctica for only peaceful purposes? It's up to us to value nature as much as we do ourselves.

ANTARCTIC INVADERS

Efforts to protect Antarctic wildlife include steps to keep invasive species from spreading across the continent. Invasive species are animals and plants that arrive in a new place from a different environment. They often spread rapidly in their new homes, prey on native species, spread diseases among native species, or capture large amounts of food, sunlight, and other resources native species depend on to survive.

For millions of years, the native species of Antarctica had no contact with other ecosystems. The plants and animals that lived there were well adapted to their icy environment. It provided the food and breeding grounds they needed to survive.

The balance started to shift when humans arrived on Antarctica. They hunted seals, whales, and other marine life, reducing the existing populations. They also brought new animals to the continent. For example, rats and mice lived aboard British sailing ships and went ashore on South Georgia and other Antarctic islands. The rodents thrived by eating seabird eggs and native grass seeds. Seabird populations plummeted.

As human exploration of Antarctica increased in the twentieth century, more and more non-native species arrived. Plant seeds and insects were carried in on the clothing and supplies of research teams. Mussels arrived on the hulls of cruise ships and reproduced in Antarctic waters. New kinds of grasses started to grow in the Antarctic soil. Some of the invaders thrived in their new homes. Others could not withstand the bitterly cold temperatures.

Conservationists knew that to protect Antarctic wildlife, the continent had to be protected from invasive species. The 1991 Protocol on Environmental Protection includes restrictions on researchers bringing non-native species to Antarctica. The protocol also prohibits dogs from the continent because they can spread a disease called distemper to seals. Conservationists have also tried to eliminate species that have already invaded. For instance, in the 2010s, a British team used poison bait to kill the rats and mice on South Georgia Island. With the rodents gone, birds will be able to flourish there again.

CLEANING AND COOKING

In 1995 biologist Carol Devine and artist Wendy Trusler, both Canadians, took part in a volunteer expedition to clean up trash left by researchers at Bellingshausen Station, a Russian (formerly Soviet) base on an island north of the Antarctic Peninsula. The trash had piled up for over twenty-eight years, and the international team of fifty-four volunteers worked all summer to collect everything from bits of plastic and old nails to fuel pipes and abandoned refrigerators. They arranged to have the trash shipped to disposal sites on other continents.

Trusler was the cook for the expedition, and the team bonded over her tasty meals. After the expedition ended, Devine and Trusler collaborated again to write *The Antarctic Book of Cooking and Cleaning: The Extraordinary Edible Record of Two Women Explorers' Journey to the End of the World*. The book tells the story of the cleanup project and also includes forty-two recipes.

The Strategic Plan for Biodiversity 2011–2020, an international project developed in 2010, is designed to protect the diversity of plant and animal life on Antarctica. All life is connected, and in Antarctica every bit of biomass can be important. Even penguin and elephant seal feces are vital to the web of life there. This waste contains nitrogen and other chemicals that nourish grasses or algae. The grasses in turn provide nesting material for birds that visit the Antarctic coast. Because of such connections, when one plant or animal species in Antarctica is harmed, others suffer as well. The Strategic Plan for Biodiversity protects wildlife on the continent through measures such as limiting fishing and research and designating areas to be protected from human activities.

In 2016 an international organization called the Commission for the Conservation of Antarctic Marine Living Resources (CCAMLR)

An orca stalks penguins from below. Pods of orcas have been observed creating waves to knock prey into the water.

created a 600,000-square-mile (1,553,993 sq. km) marine reserve in the Ross Sea that is off-limits to fishing. The protected area is a key breeding ground for whales, penguins, fish, and other marine life. Eliminating commercial fishing there will provide a refuge for these species and allow them to maintain a healthier population globally.

International protocols, plans, and reserves are important steps, but these efforts can't protect Antarctica from the effects of climate change. The Antarctic Peninsula has warmed by 4.5°F (2.5°C) since 1950, and nearly all of its glaciers have accelerated, thinned, and decreased in volume. In 1995 and 2002, large areas of floating ice on the Larsen Ice Shelf disintegrated because of increased melting. Scientists say the rate of melting continues to increase. Because of its impact on sea level, and its importance to the ecosystem of the Southern Ocean, changes in Antarctica have global consequences. Many organizations are working to make parts of the Southern Ocean a marine protected area (MPA), as was done for the Ross Sea.

CHAPTER 7

WORKING AND LIVING IN ANTARCTICA

TODAY WE ARE STILL UNDER WAY WEAVING IN AND OUT OF ICEBERGS. IT IS REALLY WONDERFUL HOW SOME ICEBERGS LOOK APPARENTLY LIKE HOUSES AND OTHER SHIPMATES SEE GROUPS OF PEOPLE ON A HILL. TONIGHT WE ARE ANCHORED AND TIED UP TO AN ICE SHELF. ON THIS LAND OF ICE, WHERE WE ARE THOUSANDS OF MILES OF ICE AND MOUNTAINS, IT'S REALLY BEAUTIFUL. SINCE OUR WATER SUPPLY IS NOT ANY TOO GOOD, TAYLOR AND I WENT OVER THE BOW OF THE [SHIP] AND JUMPED DOWN ON THE ICE. WE GOT SOME SNOW TO MELT FOR DRINKING WATER AS THE SHIP'S WATER IS RUSTY AND [HAS TO BE] STRAINED BEFORE IT IS CLEAR. IT WAS COLD GETTING SNOW, BUT I ENJOYED IT ANYHOW.

—George W. Gibbs Jr., January 25, 1940

THE SUN CAME OUT VERY BRIGHT, AS USUAL, TONIGHT AND MT. MELBOURNE (IN VICTORIA LAND) . . . WAS SIGHTED AT 11 P.M.

—George W. Gibbs Jr., February 3, 1940

Antarctica holds fifty permanent year-round bases, operated by thirty-two nations. More than thirty other bases are part-time operations, running between October and March, during the Antarctic spring and summer. During summer, the bases are home to almost four thousand scientists and staff members. In winter the number of residents falls to about twelve hundred.

Each base has a different culture and flavor, depending on the nationality of the scientists there. For example, teams at the French bases eat a daily variety of cheeses and drink wine with meals—both French culinary traditions. At the United Kingdom bases, researchers enjoy high tea, a late-afternoon tea break. That's a British tradition. At the US McMurdo base, residents sometimes eat fresh vegetables grown in a greenhouse at the base.

Argentina's Orcadas Station, which opened in 1904, is the oldest-operating base on Antarctica. Princess Elisabeth is Belgium's ecofriendly research station, powered entirely by wind and solar energy. China has four stations in Antarctica, with another one opening in 2022. One of their bases, Taishan Station, is a round, red building that resembles a traditional red Chinese lantern.

Operating under the Protocol for Environmental Protection, modern researchers take great care not to damage the areas where they work in Antarctica. They are very careful about cleaning up their trash. They reuse and recycle whatever they can and ship the leftovers back to their home countries for proper disposal. India's Bharati Station was designed to have a low impact on the land. It is made from 134 shipping containers that were prefabricated in Germany, shipped to the base, and then connected together on-site in the summer of 2011–2012. If the base ends operations, the facility can be taken apart, packed up, and shipped off without leaving any trace behind.

People living in Antarctic bases over the winter see very few green plants, unless their bases have greenhouses. The only smells are along the coast from penguins, seals, fish, and the sea. When living in the

The McMurdo Station, owned and operated by the United States, can support over twelve hundred people at once.

interior, people miss the smells of plants, flowers, and forests. The sensory deprivation can be stark, so base planners try to incorporate sights and smells into the architecture. At Russia's Vostok Station, new buildings will have cedar wood paneling, which will offer a pleasant aroma to the researchers there. They will also grow mint and other fragrant plants in indoor gardens. The South Korean base Jang Bogo on Terra Nova Bay has a large greenhouse room and a sitting area modeled after South Korean parks.

Several Antarctic bases have steam rooms or saunas. They tend to be quite popular, as they help to keep researchers warm in bitterly cold winters. At the US base at the South Pole, the Amundsen-Scott South Pole Station, the research staff go a step further. As the midwinter temperature drops to –100°F (–73°C) or below, the sauna is heated to 200°F (93°C). After spending several minutes in the superheated sauna, the adventurous staff race out into the polar night, circle the South Pole marker, and race back in. Participants are then members of the 300 Club for enduring a temperature shift of three hundred degrees, a South Pole Station winter tradition.

CUTTING-EDGE SCIENCE

Each Antarctic base runs different facilities and programs. For instance, the US-run Amundsen-Scott South Pole Station includes a project called the IceCube Neutrino Observatory. Built in the first decade of the twenty-first century, the observatory studies neutrinos, high-energy particles that travel through space. Neutrinos are like the ghosts of the atomic particle world, able to travel many miles through solid rock without being stopped. Studying them helps scientists learn about exploding stars, black holes, and other space phenomenon. To detect and measure enough neutrinos for study, scientists need giant tanks or instrumented blocks of material. These help them to see the faint, rare pulses of light emitted when a neutrino strikes an atom. To keep other particles from interfering with neutrino detection, scientists build the detectors deep underground or under the ice. The IceCube detector extends for 1.6 miles (2.5 km) under the ice, and inhabits 0.24 cubic mile (1 cu. km) of space inside the ice. In addition to studying neutrinos, IceCube also analyzes space particles called cosmic rays.

The Amundsen-Scott South Pole Station is located roughly at exactly the geographical South Pole, allowing researchers to obtain unique scientific data.

CRISIS AT THE SOUTH POLE

Jerri Nielsen, an emergency room physician from New York, took a job at the Amundsen-Scott South Pole Station in 1999. Her job was to treat station members who were sick or injured. Then Nielsen got sick herself. In May of that year, well into the polar night, she discovered a lump in her breast. She knew that the lump meant she might have breast cancer. Winter was coming to Antarctica. Temperatures were –100°F (–73°C) and lower. Airplanes can't take off or land at these temperatures, so Nielsen couldn't return to the United States to see a cancer specialist.

So Nielsen improvised with the tools at hand. She used a needle to take a tissue sample from her breast. Computer technicians at the base made digital images of the sample and sent them via satellite to doctors in the United States. They confirmed that Nielsen had cancer. She needed chemotherapy drugs, but by then Antarctica was in the dead of winter. Although planes couldn't land there, they could fly overhead and drop supplies using parachutes. That's how Nielsen got the drugs and medical equipment she needed to treat the cancer herself.

In October, spring had returned to Antarctica and Nielsen was able to fly home to the United States. There she got additional cancer treatment and wrote a book about her ordeal: *Ice Bound: A Doctor's Incredible Battle for Survival at the South Pole*. She died in 2009.

The Amundsen-Scott South Pole Station runs the South Pole Telescope. Built in 2007, the telescope collects and analyzes radio waves given off by objects in space. The South Pole is a perfect place to collect these waves because the air is extremely dry and cold. There is little water vapor in the air to interfere with incoming waves. By studying radio waves, the South Pole Telescope has helped scientists learn more

Astronomers have to use computers to convert the signals collected by radio telescopes, such as the South Pole Telescope, into images.

about the big bang—an event believed to have been the birth of the universe, which occurred about fourteen billion years ago. The South Pole Telescope was also part of the team of telescopes used in 2019 to create the first-ever picture of a black hole.

At other bases, scientists immerse themselves in geology, glaciology, weather and climate science, the study of earthquakes and volcanoes, oceanography, and dozens of other scientific pursuits. Some scientists study meteorites that have been buried in the Antarctic ice for hundreds of thousands of years. Others study the hole in the ozone layer. Research projects vary from year to year, depending on funding and government priorities.

LIFE ON THE ICE

What is the largest base on Antarctica like? McMurdo Station is known as Mactown. It's a hub for transportation, a meeting place, and a seat of power where decisions are made. More than one thousand people live there in the summer. During this season, since the sun never sets, activities take place around the clock. The residents include a staff of cooks, janitors, and "fuelies" (people who fuel airplanes and other vehicles). Other people install and repair machinery used to make measurements, drill ice cores, and run experiments. The rest of

Fieldwork in the harsh conditions is an adventure and can be quite challenging. This field toilet is encased in bricks of snow to provide privacy at a camp on Mitchell Peninsula.

the residents are mostly researchers and scientists. The average age at McMurdo is thirty-six, although some eighty-year-olds have worked there. The base rarely has anyone under twenty.

Spending time in the Antarctic summer can be pleasant, especially along the coast. Temperatures there can reach a relatively comfortable 50°F (10°C). Add in the twenty-four-hour sunshine, and an Antarctic summer at the coastal bases is fairly easy to handle. Temperatures are much lower inland, where even in midsummer the temperature can be –20°F (–29°C), but bright sunshine and lower winds help to make it tolerable. It's no surprise that most people work in Antarctica in the summer.

Spending the winter in Antarctica is much tougher. The sun doesn't rise for several months, and winds can become almost continuous. People miss the sunshine and also suffer from a lack of vitamin D, which is key to both physical and mental health. With winter temperatures dipping down to –80°F (–62°C) and colder, no one goes outside for very long. Ten minutes outside is a major excursion in these conditions. When people do go outside, they pile on layers of warm clothing, including merino wool from Australia and New Zealand as well as high-tech synthetic fibers.

Most food and supplies that base residents need must be shipped in by boat or airplane. Water is the exception. Water is plentiful in Antarctica, in the form of ice. Antarctica researchers shovel snow and ice into large tanks or stockpots and heat it to make water for drinking, bathing, and cooking. They sometimes pump liquid water from fresh Antarctic lakes during the summer. At several of the major bases, the source of potable water is the sea. The bases use a desalinization system, which employs a process called reverse osmosis to remove the salt from ocean water. Sometimes the system can't keep up with demand, and base residents must limit their water use. They take very short showers (two minutes at the most) once or twice a week. They are careful not to leave taps running when they brush their teeth and keep clothes washing to a minimum.

Louise Albershard operates a drill used for boring into ice so researchers can extract samples and take measurements.

What draws people to the Ice, as Antarctica is often called? Dr. Ted Scambos, a polar scientist who studies the behavior of the fast-flowing Thwaites Glacier, describes enjoying the challenges of living in the middle of nowhere. In Antarctica, he says, you have to make do with what you have—whether that means cobbling together a damaged piece of equipment or making a special meal out of dehydrated foods. In Antarctica, you celebrate small things, like a science experiment that goes well, finishing a short traverse to take ice measurements, or just relaxing with fellow researchers. Some scientists say that they feel much more in touch with Earth in Antarctica. The landscape is old and largely untouched by humans. It makes you want to return again and again.

VISITING ANTARCTICA

Antarctica might not sound like a prime vacation spot, but adventurous tourists have been visiting the magic land of ice, water, and light since the mid-twentieth century. In 1956 Chile's national airline took sixty-six passengers on a sightseeing flight over the South Shetland Islands, at the northern tip of the Antarctic Peninsula. The plane did not land. It simply circled around and flew back to Chile. New Zealand Air also began to take round-trip sightseeing flights to the continent. Unfortunately, one of these ended in tragedy, colliding with Mount Erebus during extreme wind conditions. After a long hiatus and a review of safety protocols, these flights and others have resumed.

Cruise ships started visiting Antarctica in the 1960s. In 1969 Swedish American adventurer Lars-Eric Lindblad organized the first commercial cruise to Antarctica on his ship *Lindblad Explorer*. Over the following decades, Antarctic tourism picked up steam. More and more companies offered sightseeing trips to the coastline. Commercial cruise ships sail to Antarctica from South Africa, Chile, and New Zealand. A few companies offer trips to the deep interior of the continent to climb its mountains or ski the last few miles to the South Pole. Needless to say, these trips are quite expensive and in general are only for the hardiest adventurers.

Tourist boats will sail along the coast of Antarctica to show visitors the breathtaking landscapes and unique wildlife such as whales and penguins.

In 1991 seven companies banded together to form the International Association of Antarctica Tour Operators (IAATO). The organization's goal is to make sure that Antarctic tourism is safe and environmentally responsible. By 2020, IAATO had about one hundred members from around the world. Almost all companies that take tourists to Antarctica are members of IAATO. They are required to comply with all international treaties and rules about caring for the continent's land, water, and wildlife. In 2019 IAATO director of operations Lisa Kelley stressed that "visiting Antarctica is a privilege and we have a responsibility to keep it pristine." The hope is that visitors continue to be ambassadors for the region and help spread the word about protecting the continent.

Antarctica hosts about sixty thousand visitors per year, most of them traveling by ship and never setting foot on the continent. Big cruise ships usually travel along the coastline, staying a safe distance from icebergs while still giving passengers breathtaking views. Some smaller vessels navigate through the ice to land on islands or on the mainland coast. Some cruises are adventure-oriented, allowing passengers to disembark and trek for short distances or kayak along ice-free coastal areas. Other tours are focused on educating visitors about Antarctic wildlife, natural history, ecology, and other science topics. They too usually stay near the coasts. Only a tiny fraction of Antarctic tourists visit the interior.

WOMEN IN ANTARCTICA

From 1914 to 1937, women from several countries applied to take part in Antarctic expeditions. They were all told no—women don't belong in Antarctica. In some cases, they were considered bad luck. During this era, most white women did not work outside the home. They were expected to take care of their husbands and children and tend to the house. Women who did work were restricted in their choice of jobs. Nursing, teaching, and secretarial work were thought to be appropriate for women. But strenuous scientific work, such as exploring the South Pole, was out of the question. Women of color were not considered at all.

In the 1930s, a few women accompanied their explorer husbands to Antarctica. These women primarily stayed with the ship and did not take part in onshore exploration. In 1947 American Finn Ronne led a private expedition to the Weddell Sea. He brought along his wife, Edith. His chief pilot Harry Darlington also brought his wife, Jennie. Edith Ronne and Jennie Darlington were the first women to "winter over," or live on the continent for a year.

In the 1950s, female scientists at last began to join Antarctic expeditions. The first was geologist Maria Klenova of the

PUTTING THE ART IN ANTARCTICA

The sights and sounds of Antarctica are awe-inspiring, and many artists convey the continent's beauty through their works. Historically, they were responsible for documenting expeditions with drawings and paintings. In modern times, many artists bring attention to melting glaciers and other ecological concerns in Antarctica.

Chicago-based artists Petra Bachmaier and Sean Gallero teamed up with geophysical sciences professor Doug MacAyeal of the University of Chicago to teach people about climate change in

Soviet Union. An Ohio State science team of six women, led by Lois Jones, were the first to the geographic South Pole in 1969. That same year, Christine Muller-Schwarze, penguin researcher, became the first woman to do research on Ross Island with her husband.

Irene Peden became the first female principal investigator (lead researcher) in 1970. Her groundbreaking research on radio waves through the ice took her to Antarctica's harsh interior. The US Navy would not allow her to be the only woman on the trip. They insisted on giving her a female assistant to accompany her. The navy also told her that if she failed, she would ruin women's chances for another generation to research on the ice.

In the following decades, more and more nations sent female researchers to Antarctic stations. This trend echoed worldwide changes in the late twentieth century. Especially in the United States and Europe, women entered scientific and other professions in large numbers. They took on jobs that had previously been held only by men. From the beginning of the twenty-first century, female researchers and other research station staff are almost as common in Antarctica as their male counterparts. However, the number of women of color in Antarctic research remains very low.

Antarctica. They created an exhibit called *White Wanderer*, which focuses on a 2,316-square-mile (6,000 sq. km) chunk of ice that broke off the Larsen C Ice Shelf into the Weddell Sea in 2017. The iceberg, which is called White Wanderer, is almost four times the size of the Hawaiian island of Oahu (the official name is far more mundane: A-68). The exhibit combines artwork and scientific data to convey the berg's enormous size and to show the dangers of climate change, melting polar ice, and rising sea levels. The artists even turned MacAyeal's seismic data—measurements of the movements and

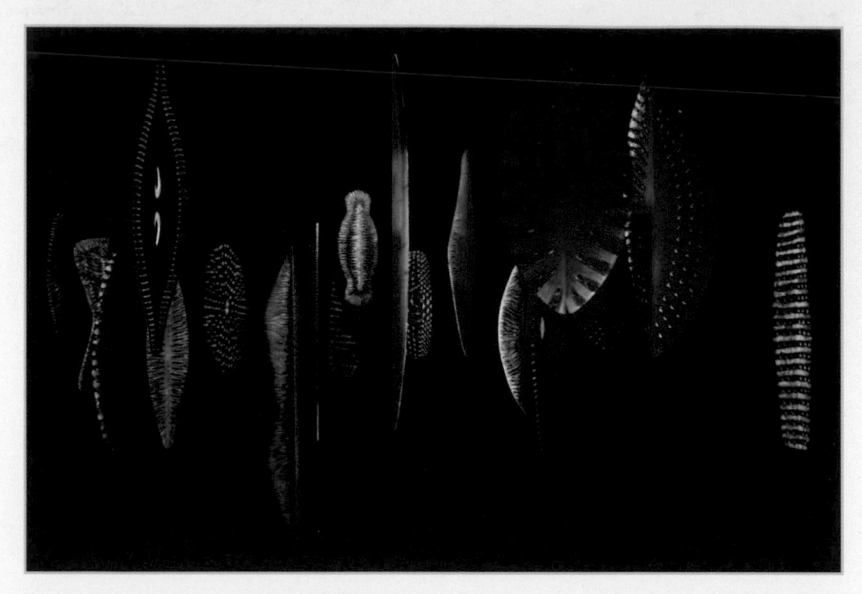

Antarctic Heart by Virginia King

cracking of the ice shelf—into sounds to convey the power and drama of the massive iceberg breaking off into the ocean.

American Paul D. Miller, who goes by the name DJ Spooky, composed a musical work that he calls Terra Nova: Sinfonioa Antarctica. This piece combines sound recordings from Antarctica with music played by a string quartet. New Zealand sculptor Virginia King was also inspired by the wonders of Antarctica. As part of an Artists to Antarctica Fellowship, she teamed up with scientists at Scott Base who were studying tiny organisms called diatoms. These microscopic algae live in the lakes and seas of Antarctica. The scientists had captured images of these creatures using electron microscopes, and King turned the pictures into dazzling colorful artwork featuring black lights and luminescent paint. She exhibited the pictures along with video footage from Antarctica and underwater recordings of the cries of Weddell seals.

Norbert Wu is a photographer and diver. He wore layers of insulated gear and braved –50°F (–46°C) temperatures to photograph the beauty that lies underwater in Antarctica. His book *Under Antarctic Ice* shows breathtaking images of penguins, jellyfish, whales, and their icy world.

This photo by Norbert Wu shows sea stars, urchins, and bootlace worms beneath ice that is attached to the sea floor.

A LOVE SONG TO THE ICE

Antarctica is a true wilderness. For scientists, it's like a living outdoor laboratory. It is complex and vast—a universe unto itself. The Ice touches one's heart in unexpected ways. The air can be too cold for words. The sun can be blinding. The wind can slap you around. But the beauty can drop you to your knees.

Many scientists and explorers have risked their lives by returning to the continent to experience its deep splendor again. The prospect of injury and death is a constant in Antarctica. Some people who travel there end up with lost fingers or toes, while others have fallen into crevasses or the sea. The silence and darkness can be lonely. Others have died from hunger or exposure. But those who traveled there say the trip is worth the risk.

Antarctica is full of questions, but it may hold some important answers. It can teach us about the origins of Earth and of the universe. It can teach us about our oceans, weather, climate, and ourselves. With its infinite shades of blue water, sunlight, and ice, Antarctica is still a place of refuge. We need the peace, stunning beauty, and solitude it can provide. In turn, the continent needs our protection. You can be a part of the creative team of scientists, artists, mechanics, and researchers who continue to find answers to the tough questions Antarctica poses for the entire planet.

AUTHOR'S NOTE

In my youth, I had little interest in what seemed like my father's distant fairy tale about a faraway icy continent. That's partly because he (George Gibbs) married my young mother, Joyce Powell, long after he had sailed to Antarctica and served in World War II. His adventures were from another time. Even though he spoke about the Antarctic voyage his entire life, I was only half listening. My friends from sixth grade remember his presentation to our class where he spoke about the explorer having to get their appendix out to be able to stay on the ice.

In spite of that, Antarctica just didn't feel relevant. When my father passed away, the writer he asked to help him complete his story, backed out. I decided it was important for me to start doing research about the expedition and his adventures. It became clear that to fully understand, I needed to take a trip to the continent. I raised the money and on February 15, 2012, I landed on King George Island, Antarctica. It happened to be Sir Ernest Shackleton's 138th birthday, which felt like a sign for me to continue the project. Antarctica, as a beacon for the world, became a passion.

During and after my exciting first encounter with the continent, I found and communicated with many Antarctic explorers, and with the descendants of the men from the US Antarctic Service Expedition of 1939–1941. I met with seamen Anthony Wayne at ninety-seven years old in his home and exchanged phone calls with Robert L. Johnson. Johnson is the last of my father's shipmates from USASE. I attended his one hundredth birthday on Zoom. Both men shared details of the expedition as though it happened yesterday. And they agreed—the continent expects the best from us in how we treat our fellow human beings.

This book honors their efforts to connect us to the continent. It is an exploration of the link between all living things to Earth, through the intersection of history, science, politics, and art of our planet's most magical place: Antarctica.

GLOSSARY

aurora: a natural display of light in the sky, usually occurring in polar regions. Auroras occur when particles from the sun hit Earth's upper atmosphere and move under the influence of its magnetic field.

crevasse: a deep crack in a glacier or the ground

echo sounding: a detection system that uses sound waves to measure the depth of water or ice

ecosystem: a community of living and nonliving things, including animals, plants, water, air, and soil, that interact with and support the health of one another

extinction: the permanent dying out of a species of plant or animal

fossil: the remains or traces of an organism that lived in the distant past. Fossils include bones that have turned to rock and the impressions that ancient organisms left in stone.

frazil: needlelike or featherlike crystals that form in seawater that is beginning to freeze

glacier: a large mass of ice that flows slowly over the land

iceberg: a piece of ice that breaks off a glacier or ice shelf

ice core: a long cylindrical piece of ice used for scientific study. Researchers use heavy machinery to drill cores and extract them from ice sheets.

ice floe: a large flat piece of floating sea ice

ice shelf: a large flat plate of ice from a glacier that extends out over the sea

invasive species: a plant or animal species that moves or is carried to a new place. Invasive species often crowd out or kill native species.

meteorite: large pieces of rock or metal that travel through space and hit Earth

nilas: thin plates of clear frozen ice on seawater

ozone hole: a region of the atmosphere where the layer of ozone gas, which protects living things from harmful rays from the sun, has thinned out. Ozone holes are formed under cold atmospheric conditions by reactions caused by human-made chemicals, which have now been mostly banned.

pack ice: a thick mass of ice formed from many ice floes gathered together

radar: a system for measuring the location of objects using radio waves

SOURCE NOTES

10 George W. Gibbs Jr., "Diary South Pole Expedition: USS *Bear* 1939–1940," last modified March 4, 1940, handwritten journal entry, December 27, 1939. In the author's possession.

11 Paul A. Carter, *Little America: Town at the End of the World* (New York: Columbia University Press, 1979), 5.

12 Carter, 6.

14 Michael Palin, *Erebus: The Story of a Ship* (London: Arrow, 2019), 104.

15, 17 Frank Wilbert Stokes, "An Artist in the Antarctic," The Antarctica Circle, accessed October 27, 2019, http://www.antarctic-circle.org/stokes.pdf.

20 S. Allen Counter, introduction to *A Negro Explorer at the North Pole: The Autobiography of Matthew Henson*, by Matthew Henson, Chicago: Independent Publishing Group May 2001, xxiii.

24 "Advert," The Antarctic Circle, accessed May 12, 2021, http://www.antarctic-circle.org/advert.htm.

26 Gibbs, "Diary," January 14, 1940.

27 Rear Admiral Richard E. Byrd, "Our Navy Explores Antarctica," *National Geographic*, October 1947, 521.

34 Gibbs, "Diary," November 22, 1939.

34 Gibbs, November 22, 1939.

35 Gibbs, November 23, 1939.

35 Gibbs, November 22, 1939.

35 Gibbs, December 3, 1939.

36 Gibbs, December 3, 1939.

35–36 Gibbs, December 19, 1939.

36–37 Gibbs, December 28, 1939.

37 Gibbs, January 10, 1940.

38 Gibbs, February 1, 1940.

38 Gibbs, February 7, 1940.

40 Gibbs, January 31, 1940.

41 Gibbs, February 23, 1940.

44 Gibbs, March 11, 1940.

46 Gibbs, March 21, 1940.

50 Gibbs, March 22, 1941.

51 Gibbs, January 10, 1941.

53 Gibbs, April 22, 1941.

54 Leilani Henry, "Antarctic Soul," Scene One, November 2018.

56 Gibbs, April 18, 1940.

56 Gibbs, April 23, 1940.

60 Gibbs, March 9, 1940.

72 Gibbs, "Diary," January 18, 1940.

74 Gibbs, January 10, 1941.

82 Gibbs, March 15, 1940.

90 Gibbs, March 7, 1940.

90 Gibbs, March 8, 1940.

106 Gibbs, January 25, 1940.

106 Gibbs, February 3, 1940.

115 Amanda Miller, "Highlights of a Polar Adventure," *EnCompass*, January/February 2019.

BIBLIOGRAPHY

Antarctic-circle.org. Accessed October 31, 2019. http://www.antarctic-circle.org.

Antarctican.org. Accessed November 1, 2019. https://www.antarctican.org/.

Behrendt, John C. *Innocents on the Ice: A Memoir of Antarctic Exploration 1957.* Niwot, CO: University Press of Colorado, 1998.

Belanger, Dian Olson. *Deep Freeze: The United States, the International Geophysical Year, and the Origins of Antarctica's Age of Science.* Boulder: University Press of Colorado, 2006.

Bertrand, Kenneth J. *Americans in Antarctica, 1775–1948.* American Geographical Society. Special publication, no. 39. New York: American Geographical Society, 1971.

Bixby, William. *Track of the* Bear*: 1873–1963.* New York: David McKay, 1965.

Bryant, Herwill. "Unpublished Antarctic Journal by Herwill Bryant." United States Antarctic Expedition 1939–1941. Unpublished manuscript, last modified April 13, 1941. PDF file.

Carter, Paul A. *Little America: Town at the End of the World.* New York: Columbia University Press, 1979.

Cool Antarctica. Accessed October 31, 2019. https://www.coolantarctica.com.

Dagnell, Lara, and Hilary Shibata. *The Japanese South Polar Expedition, 1910–12: A Record of Antarctica.* Norwich, UK: Erskine, 2011.

Gibbs, George W., Jr. "Diary South Pole Expedition: USS *Bear* 1939–1940." Handwritten journal entry, last modified March 4, 1940.

Harman, Lynn, and Christine Powell. *Knitting in Antarctica.* McMurdo, Antarctica: Green Brain, 2017.

Henson, Matthew. *A Negro Explorer at the North Pole: The Autobiography of Matthew Henson.* Montpelier, VT: Invisible Cities, 2001.

Lewis-Jones, Huw, and Kari Herbert. *Explorer's Sketchbooks: The Art of Discovery and Adventure.* San Francisco: Chronicle Books, 2016.

Lieberman, Jacob. *Light: Medicine of the Future*. Rochester, VT: Bear, 1991.

Lynch, Joseph, Jr. "A Philatelic Introduction to B.A.E. III: United States Antarctic Service Expedition 1939–1941." South-Pole.com. Accessed October 31, 2019. http://www.south-pole.com/p0000146.htm.

McArthur, Matthew Alan. *Ice Coffee: The History of Human Activity in Antarctica*. Podcasts.apple.com, 2016. https://podcasts.apple.com /pl/podcast/ice-coffee-the-history-of-human-activity-in-antarctica /id816458174?ign-mpt=uo%3D4.

Meduna, Veronika. *Secrets of the Ice: Antarctica's Clues to Climate, the Universe and the Limits of Life*. New Haven, CT: Yale University Press, 2012.

Miller, Paul D. *The Book of Ice*. Brooklyn: Mark Batty, 2011.

National Snow and Ice Data Center. Accessed October 31, 2019. https:// nsidc.org.

Richardson, H. "H. Richardson 1939–1941." Unpublished Antarctic log transcript. Last modified June 15, 1940 Microsoft Word file.

Roberts, Peder, Lize-Marié van der Watt, and Adrian Howkins, eds. *Antarctica and the Humanites*. London: Palgrave Macmillan, 2016.

Rose, Lisle. *Explorer: The Life of Richard E. Byrd*. Colombia: University of Missouri Press, 2008.

Scambos, Theodore. Interviews with the author, October 2018–June 2019.

Schwartz, Simon. *First Man: Reimagining Matthew Henson*. Minneapolis: Graphic Universe, 2015.

Stewart, John. *Antarctica: An Encyclopedia*. 2nd ed. Jefferson, NC: McFarland, 2011.

Swanson, Jennifer. *Geoengineering Earth's Climate: Resetting the Thermostat*. Minneapolis: Twenty-First Century Books, 2018.

Trusler, Wendy, and Carol Devine. *Antarctic Book of Cooking and Cleaning*. Toronto: Vauve, 2013.

Walker, Gabrielle. *Antarctica: An Intimate Portrait of a Mysterious Continent*. London: Bloomsbury, 2013.

Walker, Sally M. *Frozen Secrets: Antarctica Revealed*. Minneapolis: Carolrhoda, 2010.

FURTHER READING

Boothe, Joan N. *The Storied Ice: Exploration, Discovery, and Adventure in Antarctica's Peninsula Region*. Berkeley, CA: Regent, 2011.

Bursey, Jack. *Antarctic Night. One Man's Story of 28,224 Hours at the Bottom of the World*. Chicago: Rand McNally, 1957.

Hirsch, Rebecca E. *Climate Migrants: On the Move in a Warming World*. Minneapolis: Twenty-First Century Books, 2017.

Lerangis, Peter. *Smiler's Bones*. New York: Scholastic, 2005.

Mann, Charles C. *The Wizard and the Prophet: Two Remarkable Scientists and Their Dueling Visions to Shape Tomorrow's World*. New York: Alfred A. Knopf, 2018.

McDonough, William, and Michael Braungart. *Cradle to Cradle: Remaking the Way We Make Things*. New York: Farrar, Straus and Giroux, 2002.

Morley, Julie. *Future Sacred: The Connected Creativity of Nature*. Rochester, VT: Park Street, 2019.

Passel, Charles F. *Ice: The Antarctic Diary of Charles F. Passel*. Edited by T. H. Bauman. Lubbock: Texas Tech University Press, 1995.

Shapiro, Laurie Gwen. *The Stowaway: A Young Man's Extraordinary Adventure to Antarctica*. New York: Simon & Schuster, 2018.

FILM

Henderson, Tom. *Ice Eagles: The Story of American Aviation Triumphs and Tragedies in Antarctica*. DVD. Burlington, VT: Graceful Willow, 2018. Part One is about the history of exploration. Part Two focuses on the role aviation plays in Antarctic science. Outside of military combat, the most dangerous environment for flying is Antarctica. This documentary interviews experienced explorers and researchers.

Young, Peter. *The Last Ocean: The Toothfish and the Battle for Antarctica's Soul*. DVD. Christchurch, NZ: Fisheye Films, 2012. This documentary explains the reason the Ross Sea needs protection. People worldwide came together to ensure the most important marine ecosystem on Earth remains unspoiled.

INDEX

PHOTO ACKNOWLEDGMENTS

Image credits: Backgrounds: jo Crebbin/Shutterstock.comVolodymyr Goinyk/Shutterstock.com, p. 7; © Ted Scambos, pp. 8, 12, 45, 48, 52, 67, 75 (all) 80, 113; Courtesy of Leilani Henry, pp. 9, 17, 27, 31, 35, 40, 41, 47, 58–59, 120; Philipp Salveter/Shutterstock.com, p. 15; Laura Westlund/Independent Picture Service, pp. 16, 62, 81, 93, 101; © Jim Richardson, pp. 18, 72; Bettmann/Getty Images, p. 20; New York Times Co./Hulton Archive/Getty Images, p. 23; © Lucia deLeiris, *The James Caird*, oil on wood panel, 9" x 12", p. 29; © Laguna Art Museum/Leland Curtis, p. 28; Pictorial Press Ltd/Alamy Stock Photo, p. 32; Michael Van Woert, NOAA NESDIS, ORA Caption/Wikimedia Commons (Public Domain), p. 36; Courtesy of the artist and Winston Wächter Fine Art p. 39; Courtesy of Ted Scambos via Ron Jesseberg Collection/National Archives, p. 43; Courtesy of American Geographical Society Library, p. 49; AP Photo/ANN BANCROFT, p. 55; © Kathleen Hill Rackley, p. 57; Gonzalo Solari Cooke/Shutterstock.com, p. 65; Colin Harris/era-images/Alamy Stock Photo, p. 66; Denis Burdin/Shutterstock.com, p. 69; Ion Mes/Shutterstock.com, p. 76; Brian L Stetson/Shutterstock.com, p. 77; Jeff Warneck/Shutterstock.com, p. 78; © 2018 Kirsten Carlson, Fathom It Studios, p. 79; Jason Edwards/Alamy Stock Photo, p. 83; Science Photo Library/Alamy Stock Photo, p. 84; JordiStock/Shutterstock.com, p. 86; Arctium Lappa/Shutterstock.com, p. 88; Zaruba Ondrej/Shutterstock.com, p. 89; Gonzalo Solari Cooke/Shutterstock.com, p. 91; ccpixx photography/Shutterstock.com, p. 95; SannePhoto/Shutterstock.com, p. 99; Luis Davilla/Getty Images, p. 105; Kevin OConnell Photography/Alamy Stock Photo, p. 108; Mesa Studios/Shutterstock.com, p. 109; Viktor Barricklow, p. 111; Marco Ramerini/Shutterstock.com, p. 115; © Virginia King, Antarctic Heart 1999–2000 www.virginiakingsculptor.com, p. 118; © #WED0040, 2021 Norbert Wu/Norbert Wu Productions, p. 119.

Cover: jo Crebbin/Shutterstock.com.